AF749332

Plant Physiology Annual Volume 2023

Edited by Jen-Tsung Chen

Published in London, United Kingdom

Plant Physiology Annual Volume 2023
http://dx.doi.org/10.5772/intechopen.113980
Edited by Jen-Tsung Chen

Contributors
Afsal Manekkathodi, Kelath Murali Manoj, Yanyou Wu, Nikolai Bazhin, Abhinav Parashar, Arturo Solís Herrera, María del Carmen Arias Esparza, Renu Kathpalia, Anita Kamra Verma, Xiaoyan Liu, Ryushiro D. Kasahara, Moslem Bahmankar, Maryam Salehi, Mohammad Reza Naghavi, Saikat Biswas, Rupa Das, Lay Lay Nwe, Elena Deineko, Ilya Zlobin, Marina Efimova, Natalya Permyakova, Irina Sokolova, Vladimir Kuznetsov

© The Editor(s) and the Author(s) 2023
The rights of the editor(s) and the author(s) have been asserted in accordance with the Copyright, Designs and Patents Act 1988. All rights to the book as a whole are reserved by INTECHOPEN LIMITED. The book as a whole (compilation) cannot be reproduced, distributed or used for commercial or non-commercial purposes without INTECHOPEN LIMITED's written permission. Enquiries concerning the use of the book should be directed to INTECHOPEN LIMITED rights and permissions department (permissions@intechopen.com).
Violations are liable to prosecution under the governing Copyright Law.

Individual chapters of this publication are distributed under the terms of the Creative Commons Attribution 3.0 Unported License which permits commercial use, distribution and reproduction of the individual chapters, provided the original author(s) and source publication are appropriately acknowledged. If so indicated, certain images may not be included under the Creative Commons license. In such cases users will need to obtain permission from the license holder to reproduce the material. More details and guidelines concerning content reuse and adaptation can be found at http://www.intechopen.com/copyright-policy.html.

Notice
Statements and opinions expressed in the chapters are these of the individual contributors and not necessarily those of the editors or publisher. No responsibility is accepted for the accuracy of information contained in the published chapters. The publisher assumes no responsibility for any damage or injury to persons or property arising out of the use of any materials, instructions, methods or ideas contained in the book.

First published in London, United Kingdom, 2023 by IntechOpen
IntechOpen is the global imprint of INTECHOPEN LIMITED, registered in England and Wales, registration number: 11086078, 5 Princes Gate Court, London, SW7 2QJ, United Kingdom

British Library Cataloguing-in-Publication Data
A catalogue record for this book is available from the British Library

Additional hard and PDF copies can be obtained from orders@intechopen.com

Plant Physiology Annual Volume 2023
Edited by Jen-Tsung Chen
p. cm.

This title is part of the Physiology Book Series, Volume 23
Series Editor: Tomasz Brzozowski

Print ISBN 978-0-85014-519-9
Online ISBN 978-0-85014-520-5
eBook (PDF) ISBN 978-0-85014-521-2
ISSN 2631-8261

We are IntechOpen, the world's leading publisher of Open Access books

Built by scientists, for scientists

6,700+
Open access books available

181,000+
International authors and editors

195M+
Downloads

156
Countries delivered to

Our authors are among the
Top 1%
most cited scientists

12.2%
Contributors from top 500 universities

WEB OF SCIENCE™

Selection of our books indexed in the Book Citation Index in Web of Science™ Core Collection (BKCI)

Interested in publishing with us?
Contact book.department@intechopen.com

Numbers displayed above are based on latest data collected.
For more information visit www.intechopen.com

For EU product safety concerns:
IN TECH d.o.o., Prolaz Marije Krucifikse Kozulić 3, 51000 Rijeka, Croatia,
info@intechopen.com or visit our website at intechopen.com.

IntechOpen Book Series
Physiology
Volume 23

Aims and Scope of the Series

Modern physiology requires a comprehensive understanding of the integration of tissues and organs throughout the mammalian body, including the cooperation between structure and function at the cellular and molecular levels governed by gene and protein expression. While a daunting task, learning is facilitated by identifying common and effective signaling pathways mediated by a variety of factors employed by nature to preserve and sustain homeostatic life. As a leading example, the cellular interaction between intracellular concentration of Ca+2 increases, and changes in plasma membrane potential is integral for coordinating blood flow, governing the exocytosis of neurotransmitters, and modulating gene expression and cell effector secretory functions. Furthermore, in this manner, understanding the systemic interaction between the cardiovascular and nervous systems has become more important than ever as human populations' life prolongation, aging and mechanisms of cellular oxidative signaling are utilised for sustaining life. Altogether, physiological research enables our identification of distinct and precise points of transition from health to the development of multimorbidity throughout the inevitable aging disorders (e.g., diabetes, hypertension, chronic kidney disease, heart failure, peptic ulcer, inflammatory bowel disease, age-related macular degeneration, cancer). With consideration of all organ systems (e.g., brain, heart, lung, gut, skeletal and smooth muscle, liver, pancreas, kidney, eye) and the interactions thereof, this Physiology Series will address the goals of resolving (1) Aging physiology and chronic disease progression (2) Examination of key cellular pathways as they relate to calcium, oxidative stress, and electrical signaling, and (3) how changes in plasma membrane produced by lipid peroxidation products can affect aging physiology, covering new research in the area of cell, human, plant and animal physiology.

Meet the Series Editor

Prof. Dr. Thomas Brzozowski works as a professor of Human Physiology and is currently a Chairman at the Department of Physiology and is V-Dean of the Medical Faculty at Jagiellonian University Medical College, Cracow, Poland. His primary area of interest is physiology and pathophysiology of the gastrointestinal (GI) tract, with a major focus on the mechanism of GI mucosal defense, protection, and ulcer healing. He was a postdoctoral NIH fellow at the University of California and the Gastroenterology VA Medical Center, Irvine, Long Beach, CA, USA, and at the Gastroenterology Clinics Erlangen-Nuremberg and Munster in Germany. He has published 290 original articles in some of the most prestigious scientific journals and seven book chapters on the pathophysiology of the GI tract, gastroprotection, ulcer healing, drug therapy of peptic ulcers, hormonal regulation of the gut, and inflammatory bowel disease.

Meet the Topic Editor

Jen-Tsung Chen is a professor of cell biology at the National University of Kaohsiung in Taiwan. He teaches genomics, proteomics, plant physiology, and plant biotechnology. Dr. Chen's research interests also include bioactive compounds, chromatography techniques, plant molecular biology, plant biotechnology, bioinformatics, and systems pharmacology. He is an active editor of academic books and journals with the aim to advance the exploration of multidisciplinary knowledge involving plant physiology, plant biotechnology, nanotechnology, ethnopharmacology, and systems biology. He serves as an editorial board member and a guest editor in reputed journals. Dr. Chen published books in collaboration with international publishers, and he is currently handling book projects for several international publishers on diverse topics. He has been included in the World's Top 2% Scientists 2023 by Stanford University.

Contents

Preface

Over the decades, the entire world has faced critical crises, primarily due to the expanding human population resulting in environmental pollution, food shortages, and the potential exacerbation of these issues under the scenario of global climate change. Scientists are expected to address these problems by exploring knowledge in plant physiology, with a specific focus on stress-tolerant mechanisms. This exploration can significantly advance agricultural biotechnology and breeding, specifically concentrating on improving crop resilience to climate change. The goal is to enhance both the yield and quality of crops, contributing to global food security. Therefore, there is a growing demand to expand our understanding of developing climate-resilient crops.

This volume offers updated methods and current applications of genome editing technology, outlining strategies to enhance drought tolerance in crops. The underlying mechanism centralizes modified cell signaling through precision mutagenesis introduced by CRISPR/Cas-mediated genome editing. CRISPR-edited plants exhibit a re-programmed balance and interconnections of plant hormones, including abscisic acid and cytokinins, attributed to edited versions of transcription factors and cytokinin oxidase/dehydrogenases.

An alternative solution to address climate-related abiotic stresses is explored through eco-friendly organic farming approaches. This volume provides a comprehensive summary of organic farming systems to alleviate abiotic stress responses in crops. It covers the application of bio-fertilizers, bio-priming with plant growth-promoting microorganisms, bio-stimulants, and organic manures. Additionally, smart agronomic practices under stress conditions, such as crop rotation, grafting, intercropping, selection of resistant/tolerant varieties, sowing methods and timing, tillage, and the use of organic formulations, are discussed.

In the face of the modern world's energy shortage, this book explores promising solutions, with a focus on exploiting alternative energy sources. It summarizes current achievements in green-energy technology, specifically artificial photosynthesis systems. These systems have the potential to increase clean energy sources like carbon monoxide, formic acid, methane, and pure hydrogen by designing an advanced photosynthetic system. The book not only illustrates mechanistic insights into photosynthetic chemistry but also comprehensively discusses the limitations and future directions of applied photosynthesis approaches. Despite these limitations, scientists are actively working to improve artificial photosynthesis systems, aiming for future technology that is more cost-effective, robust, and scalable.

In this volume, various interesting subtopics of plant physiology and cross-field observations are proposed. For instance, a biochemistry perspective on the dissociation of water to produce oxygen in plants is presented, with the intention of inspiring readers to design future experiments for further clarification. Additionally, a discussion on the potential role of a 'robber tree' and insights into plant reproductive strategies is put forth, suggesting their potential applications in agriculture in the future.

This book is an ideal reference for students, teachers, and scientists in the field of plant science. Specifically, researchers focusing on subtopics of plant stress physiology, plant biochemistry, photosynthesis, and crop breeding are highly recommended.

Jen-Tsung Chen
Department of Life Sciences,
National University of Kaohsiung,
Kaohsiung, Taiwan

Chapter 1

The Modification of Abscisic Acid and Cytokinin Signaling with Genome Editing to Increase Plant Drought Tolerance

Ilya Zlobin, Marina Efimova, Natalia Permykova, Irina Sokolova, Vladimir Kuznetsov and Elena Deineko

Abstract

Due to climate aridization, the need to increase the resilience of plant productivity lo water stress becomes urgent. Abscisic acid and cytokinins have opposing biological roles during water deficit and post-drought recovery, but both these regulators can be utilized to maintain plant productivity under water stress. Downregulation of abscisic acid biosynthesis and signaling can aid in the maintenance of photosynthesis, growth, and productivity in plants, although increasing the susceptibility to severe stress. Cytokinin upregulation can maintain photosynthesis and productivity during water stress and aid recovery processes, whereas downregulation can lead to increased root growth, thus improving plant water balance, nutrient absorption, and hence productivity in water-limited conditions. The use of modern genome editing methods makes it possible to specifically modify genes involved in the implementation of complex traits in plants, such as resistance to stress factors. This review will examine the main areas of work on genome editing of gene families involved in plant responses to water deficiency using CRISPR/Cas technologies. Our current work on editing the ABF gene family, encoding transcription factors for ABA (AREB1/ABF2, AREB2/ABF4, and ABF3), as well as the CKX gene family (CKX1 and CKX4), encoding cytokinin oxidase/dehydrogenases, will be presented.

Keywords: stress factors, plant drought resistance, molecular methods, plant genome modification, genome editing, abscisic acid, cytokinin signaling

1. Introduction

Drought is the most important abiotic factor challenging plant survival, performance, and productivity on the planet. The rapidly increasing risk of coupled negative effects of water deficit and heat stress implies that we need to adapt the physiology of major crop plants to a hotter and drier future. Generally, the adaptation of annual crop plant to water stress can be confined to two capabilities:

 IntechOpen

- to survive water stress without major irreversible effects on plant performance
- to quickly and fully recover after restoration of water supply, thus decreasing the cumulative drought impact on plant performance

It is clear that these capabilities depend on different or even conflicting plant traits. Survival during water deficiency depends mainly on the ability to prevent the irreversible desiccation of plant tissues and maintain the hydraulic integrity of a plant [1]. Plants need to minimize water losses, whereas photosynthesis and growth can be drastically diminished during this period without the major threat for plant survival during drought. In contrast, recovery occurs when water is plenty again and depends on the ability to recover photosynthesis, growth, and resource allocation to productive organs. Therefore, it is likely that opposing regulatory mechanisms would be required to make plants more tolerant to water stress per se and to make them abler to recover form water stress [2]. A clear example of such opposing pair of regulatory mechanisms are abscisic acid (ABA) and cytokinins (CKs). This chapter is devoted to the effects of these regulatory molecules on plant performance during water stress and recovery, on their regulatory modes, and on the usage of genome editing technologies to change plant ABA and CK balance to increase drought tolerance and post-drought recovery.

2. Abscisic acid

The response to water deficit is the major biological function of abscisic acid, and ABA can be considered as a versatile hormone that regulates plant water status in an integrated fashion. Abscisic acid increases water acquisition by affecting root growth and plant osmotic balance, affects water transport from the root surface to leaf tissues through regulation of aquaporin genes, and regulates water spending through the regulation of stomatal conductance, possibly influencing cuticular conductance. In case of stress severe enough to exert a substantial degree of dehydration in plant cells, ABA regulates the biosynthesis of stress-protective compounds such as dehydrins [3], but such intensive stress is likely of minor importance for agricultural plants [4]. The major biological functions of ABA during water stress are considered below.

2.1 Abscisic acid and plant water spending

The most-studied biological effect of ABA is stomatal closure, which enables plants to greatly diminish water losses, thus making a major contribution to the maintenance of plant water status during drought [5]. The paramount importance of ABA for stomatal regulation is clearly illustrated by ABA-deficient mutants, which are extremely sensitive to increasing vapor pressure deficit even under well-watered conditions [6]. In more ancient plant lineages, ABA biosynthesis is rather slow, likely due to a reliance on non-specific enzymes during ABA biosynthesis, and therefore, ABA accumulation can occur only if water stress is rather prolonged [7–9]. For example, in [10], ABA increment in drying leaf tissues in several coniferous species occurred only after 2-h of dehydration. In contrast, angiosperms can induce ABA biosynthesis rapidly within few tens of minutes through the activation of NCED gene expression in ABA-producing tissues [7]. ABA catabolism genes are often upregulated simultaneously with ABA biosynthesis genes, but the expression of the former is lower than the latter, resulting in net ABA accumulation under inductive conditions [11]. ABA accumulation in dehydrating cells occurs due to the decreasing of cell volume, rather than of turgor or water potential [12]. After water stress relief, the activity of ABA biosynthesis genes stayed elevated in guard cells, thus achieving drought memory effects favorable in case of subsequent droughts [13]. ABA controls memory processes through signaling pathway SnRK2/ABF/ABRE [14]. ABA biosynthesis in roots also occurs under water stress, but leaf is likely a major site of ABA biosynthesis [15, 16]. There are conflicting evidences whether leaf ABA biosynthesis occurs mainly in vascular buds and guard cells or in mesophyll tissues, with solid evidences in favor of mesophyll as the main site of ABA biosynthesis [7, 10, 15]. Additionally, in angiosperms, rapid ABA-induced stomatal closure occurs within seconds to minutes, thus indicating the presence of ABA-dependent non-transcriptional mechanisms in stomatal closure [17]. Abscisic acid is among the key mechanisms underlying the difference between isohydric (R-type) and anisohydric (P-type) strategies under drought stress, as isohydric plants achieve high leaf ABA levels during stress, whereas anisohydric plants respond to stress with an initial peak and subsequent decline of ABA content [5, 18, 19]. In ferns or lycophytes, the stomatal closure is independent of ABA [5], thus indicating that stomatal regulation by ABA is a relatively late evolutionary achievement.

Besides regulation of stomatal conductance, ABA probably plays a role in the regulation of residual cuticular conductance. This type of conductance, although quantitatively minor in well-watered plants, becomes the major determinant of plant survival during prolonged water deficiency, when water absorption by the plant root system reaches zero, and the ability to preserve the water already present in tissues becomes crucial [1]. Cuticle is often viewed as a rather stable structure hard to be modified, but in fact, recent assimilates can be incorporated rapidly in the cuticle [20], indicating that cuticle can be probably abler to modifications than it is thought currently. Plant minimum leaf conductance can decrease under drought stress from −4 to −70%, with a decrease of 30–40% being typical [21]. It is known that ABA can change the chemical composition of cuticle, but whether it aids in decreasing minimal conductance is unknown and requires further clarification [21].

2.2 Abscisic acid and regulation of water acquisition and transport

ABA effect on root growth is biphasic, with mild ABA increase stimulating root growth through ethylene-dependent mechanisms, whereas higher ABA concentrations inhibit growth through auxin signaling pathway [22, 23]. The ABA-dependent increase in main root elongation concomitantly with the inhibition of lateral root formation aids plants in reaching deep water-containing soil horizons with minimal carbon expenditure on root growth, and the biological effects of ABA and drought on root growth are similar [24]. However, in [25], ABA increased lateral root number and length at mild water deficit, likely also suppressing primary root growth to a certain extent. Not only the biological effects of ABA but also the source of ABA in root remain somewhat controversial. Mild drought leads to local ABA accumulation in roots [26]. Earlier, root tip was thought to be the main source of ABA biosynthesis during water stress, but now, it is clear that leaf-derived ABA plays a major role in shaping root growth [27], whereas ABA biosynthesis in roots can be limited by carotenoid substrate limitation under water stress [10]. ABA effects on root growth take place via interacting network with cytokinins, ethylene, and auxin [23]. Synthesis of ABA in roots of transgenic poplar increased root growth and drought tolerance [28].

ABA is involved in the stimulation of reversible suberization of root endodermis, which is required to regulate the apoplastic movement of water [17]. The decrease in root hydraulic conductance, in turn, leads to stomatal closure and water economy, aiding in adaptation to water deficiency [29]. The decrease in ABA accumulation in root tissues during water stress can be also observed, probably due to increased ABA translocation to the above-ground plant part [30]. In addition to root growth, ABA can also positively affect osmotic adjustment, thus increasing plant water absorbing capacity [31].

Abscisic acid participates in regulation of the plant aquaporin system [32]. ABA positively affects root hydraulics [25] and is involved in jasmonate-mediated increase of root hydraulic conductivity [33]. However, not only promoting but also inhibiting effects of ABA on root hydraulic conductivity are observed [26]. Also, ABA can play a role in increasing the water-transporting ability of mycorrhizal fungi [26]. ABA is among the key regulators of expression of aquaporin genes [26]. In Zea mays, ABA increases both gene expression and protein content of different PIP aquaporins, although the results can vary between studies [34]. Also, ABA participates in the regulation of aquaporin activity through phosphorylation [32]. Although ABA is generally viewed as hormone inhibiting the above-ground growth, the positive ABA influence on plant hydraulic conductance through the regulation of aquaporin system can translate into positive effect on leaf extension growth, thus making the total ABA effect on growth less straightforward [34]. ABA-induced decrease of leaf hydraulic conductivity can participate in the regulation of stomatal closure [35].

The role of ABA in the regulation of axial water transport through xylem is less well studied, compared to cell-to-cell transport through aquaporins. ABA is well-known to regulate the blockage of plasmodesmata in dormant cambium, making it unresponsive to activating environmental signals, and is involved in the termination of wood differentiation [36]. Exogenous ABA treatment often leads to reduced stem growth through inhibition of cambial activity, whereas occasional reports of secondary growth stimulation by ABA treatment likely stem from specific experimental approach rather than from ABA effects per se [37]. In cambial and xylem tissues of *Eucommia ulmoides* trees, the seasonal dynamics of ABA and IAA was the opposite, and ABA negatively influenced cambium reactivation by IAA [38]. ABA treatment decreases the hydraulic diameter of vessels, which negatively affects xylem hydraulic conductance [39]. Therefore, ABA likely plays a negative role in the formation of water-transporting tissues during plant secondary growth.

2.3 Trade-offs of ABA effects on plant performance

The above-described integrative positive effects of ABA on plant drought tolerance are linked with several important trade-offs. Although ABA biosynthesis is down-regulated quite rapidly during post-stress period, the major increment of ABA in leaves can sustain for prolonged period after drought release [5]. This limits a plant's ability to rapidly restore gas exchange and photosynthesis and underlies, at least partially, the hysteresis between stomatal conductance and other leaf hydraulic characteristics post-drought [40], although these limitations can be also unrelated to ABA accumulation. However, it should be noted that sustained ABA accumulation may aid the recovery processes by facilitation of embolism repair by decreasing stomatal conductance and water loss, which favors embolism refilling processes [41, 42]. Also, memory effects due to ABA increase during the first stress encounter can increase the tolerance to subsequent stresses and yield [43]. The negative influence of ABA on leaf growth can be mainly due to the inhibition of assimilation resulting in source limitation of growth [10]. However, direct negative ABA effects on growth processes through ABF transcription factors is also well-known [44]. The ABA-induced increase of biosynthesis of osmolytes and protective compounds would distract these resources from growth and reproduction. Also, allocation of belowground growth to the deeper root system would probably lead to deterioration of mineral nutrition, since the deeper soil layers are deprived with mineral nutrients compared to upper layers [45].

Given these trade-offs, it is not surprising that constitutively increased ABA biosynthesis and ABA signaling results in depressed growth and productivity in non-stressed conditions, whereas the suppression of ABA signaling increases growth in the absence of abiotic stressors [46]. Crop plants are usually grown in more favorable conditions compared to native plants, and severe water stress is less prevalent for agricultural ecosystems compared to native ones [4]. Also, the maintenance of productivity during mild water stress is obviously more important from the economical point of view compared to the ability to survive severe water deficiency, since in the latter case, the productivity would be anyway lost. Therefore, for annual crops, the downregulation rather than upregulation of ABA biosynthesis and/or signaling can be a more promising strategy to maintain productivity during mild stress, although making plants more susceptible to severe stresses, which are devastating for plant productivity no matter whether plants survive the stress period or are desiccated. However, it is known that the logarithmic character of dependence of carbon fixation on stomatal conductance means that plants can decrease stomatal conductance to a certain extent without trade-off with CO2 uptake and assimilation activity [5]. It can be therefore proposed that mild increase in ABA biosynthesis/signaling with the associated moderate decrease of stomatal conductance can result in substantially improved water use efficiency without compromising plant productivity, making such plants more effective from the economical point of view.

3. Cytokinins and their effects on plant performance during drought and recovery

Generally, biosynthesis and signaling of cytokinins are negatively affected by drought, consistent with the view on CKs as negative regulators of drought tolerance [11]. However, the regulation of CK metabolism under water-stress conditions can be rather specific, with different IPT genes demonstrating differently directed regulation under drought, whereas for CK OXIDASES/DEHYDROGENASES (CKX), more uniform upregulation is observed [47]. The directional changes in CK biosynthesis and signaling can have rather contrasting effects on plant ability to tolerate drought and to recover from its impact. Both CK signaling mutants and transgenic plants with enhanced CK signaling often demonstrate increased drought tolerance (Hai 2020). CKs generally exacerbate water loss by plants, thus making them more prominent to severe drought, whereas decreased CK levels contribute to more parsimonious water spending and better maintenance of plant water status during stress [48]. Also, CKs are positive regulators of shoot meristem activity and hence shoot growth [11, 49], and the increased above-ground growth can be maladaptive under severe water deficiency. The decreased CK accumulation can be associated with higher tolerance of photosynthetic processes during drought [48]. CKs and ABA reciprocally downregulate the biosynthesis and signaling of each other, thus exerting contrasting effects in plants under non-stressed conditions and under drought stress [50]. ABA decreases CK contents, which increase plant sensitivity to ABA, thus making plants abler to respond to water deficiency [50]. Cytokinins repress SnRKs as major components of ABA signaling, thus inhibiting ABA effects on plant under non-stressed conditions [50].

On the other hand, when water stress is not severe and plants are not at risk of desiccation, CKs can have numerous positive effects on plant drought and post-drought performance. Both exogenous CK treatment and modulation of endogenous CK levels were reported to positively affect plant drought tolerance [48]. Increased CK biosynthesis delayed drought-induced leaf senescence in tobacco and maintained photosynthesis, thus decreasing yield loss [4]. Under water deficiency, CKs promote stomatal conductance and chlorophyll biosynthesis [47], which can be detrimental under severe stress but is advantageous for productivity under relatively mild water stress conditions. CK-mediated inhibition of stomatal closure is the conserved response in diverse plant species [5]. CKs promote plant antioxidant defense by increasing the activity of antioxidant enzymes and decreasing the activity of ROS-generating systems such as xanthine oxidase [47]. CKs may positively affect plant osmotic adjustment under water deficiency [51]; in contrary, in [4], much lower proline accumulation in tobacco plants with increased CK biosynthesis was observed, likely due to their higher drought tolerance and lower degree of stress compared to wild-type plants. Also, CKs positively influence cambial activity and radial growth, thus increasing stem hydraulic conductance [37]. Generally, many of the positive CK effects during mild stress can be due to a delay in activation of drought response, thus decreasing stress impact [48]; CKs are known to suppress SnRK2 functioning and thus stress response [50]. It can be very promising for agricultural plants, since plant productivity and not plant survival is of the most interest for agriculture, thus making CK-induced desensitization of plants to environmental stress a promising strategy to maintain crop productivity under relatively mild stresses, typical for agricultural conditions [4].

CKs have numerous positive effects on plant post-drought recovery processes, making plants with upregulated CK content superior in recovery compared to wild-type plants. CKs in plants decreases under drought [11] while increasing prominently during the recovery period, together with compensatory growth acceleration compared to non-stressed plants [48]. Higher CK content during the post-drought recovery period can elevate auxin content in leaves [48] and also in cambium [52], which is necessary for active post-drought growth. CKs positively affect stomatal opening in post-drought period [53], helping to restore photosynthesis and to minimize cumulative negative drought impact on assimilation. Therefore, despite negative effects of CKs on tolerance to severe stress, their upregulation can be a promising way to increase the performance and productivity of crop plants.

However, downregulation of CKs can also have positive effects on crop performance under water shortage. Cytokinins are negative regulators of root meristem activity [11, 54], suppressing both primary root elongation and root branching [50, 55]. Negative CK effect on primary root growth is exerted through increase in ethylene biosynthesis [54]. As a result, CKs decrease both drought tolerance and absorption of mineral nutrients [56], whereas reduction of endogenous CKs can have prominent positive effects on root growth, increasing the number and length of lateral roots and root biomass accumulation [47]. For example, root-specific expression of CKX gene in Zea mays improved both root growth and mineral nutrition of plants, which was surprisingly achieved without trade-offs with above-ground growth [57].

The fact that such prominent changes in whole-plant architecture were made possible by expressing a single gene is quite promising for plant improvement. Therefore, both decrease and increase in CK biosynthesis and signaling can be viewed as a potential way to increase the resilience of crop productivity to water shortage.

4. Methods for modifying plant genomes

Plant genetic traits are inherited from parents from generation to generation and are encoded by genetic information contained in DNA. At the same time, genetic information is subject to constant changes due to the presence of spontaneous or induced mutations, errors arising during transcription, the activity of transposable elements, the processes of meiotic crossing over, and cross-fertilization. Some pathogenic and symbiotic bacteria, such as Agrobacterium spp. [58], can transfer part of their DNA into the genome of the host cell, thereby changing the functioning of the host cell to suit their needs. Thus, genome modification occurs constantly in a plant cell.

Plant breeding is the process of obtaining new varieties of plants that contain in their genome a set of genes that make it possible to grow plants that are suitable for agricultural production, processing, and consumption and at the same time have properties beneficial to humans and animals. Thus, plant breeding involves systematic selection among the entire population of plants of samples bearing target properties. It is estimated that humans have been successfully breeding plants for over ten thousand years [59] when seeds of plants with favorable features were saved for the next plantation, a practice known as domestication. The most significant advances in plant breeding techniques have been achieved as knowledge and understanding of plants and their genetic structures have accumulated. In the second half of the twentieth century, with an increase in the quantity and quality of food consumption, a revolution in plant breeding occurred, the key achievements of which were achieved in the creation of hybrids and transgenesis. The most important stage in plant breeding was the Green Revolution, which made it possible to dramatically increase the productivity of agricultural crops through the development of high-yielding varieties of cereals, particularly dwarf wheat and rice. Norman Borlaug, Nobel Prize laureate and father of the Green Revolution, emphasized that the key to the success of these semi-dwarf varieties was their wide adaptability, short plant height, high sensitivity to fertilizers, and resistance to disease, which ultimately made it possible to obtain more yield at a lower cost [59]. Later, these requests were addressed to the emerging technology of transgenesis, which led to its rapid development. Transgenic crops are now widespread globally and are increasingly accepted as food and feed. Transgenesis changes the genetic information of a plant cell, resulting in a so-called genetically modified organism (GMO) that carries in its genome a fragment of foreign DNA that gives the plant new useful traits that cannot be obtained by conventional breeding methods. However, GMO organisms were perceived ambiguously by society, which led to the fact that obtaining state registration for a GMO variety in some countries is significantly difficult or completely impossible.

4.1 Development of the genome editing tools

With the development of genetic engineering methods and the accumulation of data on plant genomes, gene editing technologies began to develop—making it possible to make site-specific changes in the target site of the genome. The first methods that appeared were zinc-finger nuclease (ZFN) and later transcription activator-like effector nucleases (TALEN). Both TALEN and ZFN are composed of repeated tandem sequences of DNA-binding domains and an attached Fok1 nuclease protein, such that the recombinant protein can be targeted to recognize a target DNA sequence and therefore create double-strand breaks (DSBs) at the target site. For each target site, a new TALEN or ZFN protein must be prepared to recognize the target DNA sequence, which required labor-intensive genetic engineering and significantly limited the widespread use of these gene editing technologies [60, 61]. However, there are examples of successful use of ZFN to manipulate genes in tobacco, Arabidopsis, and maize [62–64]. TALENs, which are easier to target to a specific DNA region because each TALEN domain recognizes one target nucleotide, as opposed to ZFN, where each domain recognizes a triplet of nucleotides, have been successfully used in horticultural crops such as soybeans, wheat, rice, tomatoes, and potatoes [65, 66]. However, the major drawback related to ZFNs and TALENs are their off-targeting effects, prolonged screening process, toxicity to the host cell, and complex genetic engineering procedures, limiting their applicability. The most modern method of genome editing is CRISPR technology; the first article on the successful application of this technology on plant cells was published in 2013, and the first edited plants were *Arabidopsis thaliana* and *Nicotiana benthamiana* [67].

Typically, CRISPR/Cas9 is a complex consisting of two components: the Cas9 endonuclease protein and a single guide RNA (sgRNA) with 20-nucleotide homology to the target DNA region [68–70]. The Cas9 endonuclease binds to the protospacer adjacent motif (PAM) DNA sequence (for Cas9 the PAM site is NGG), the sgRNA complementarily binds to the DNA sequence adjacent to the PAM site, and if the binding is successful, Cas9 carries out a DSB in the target site [68, 71]. DSBs caused by the Cas9 endonuclease lead to the activation of DNA repair systems, which can take two pathways, the error-prone non-homologous end-joining (NHEJ) or homology-directed repair (HDR). Errors of DNA repair system result in deletions, insertions, or substitutions of DNA at DSB sites, which in turn disrupt gene function or cause a reading frameshift, known as a gene mutation or knockout [68–70]. As a result of DSB repair via the NHEJ pathway, insertions/deletions (indels) of several bases are usually observed during plant genome editing based on CRISPR/Cas9. The use of the mechanism of HDR, in turn, makes it possible, using editing systems, to replace individual nucleotides in the DNA sequence and even obtain a site-specific insertion of a gene or group of genes.

At the moment, editing technologies have become so widely developed that they make it possible to influence any stage of the implementation of genetic information in a cell—at the level of transcription, translation, post-translation, epigenetic and so on [72]. Over the past 10 years, a number of different CRISPR-based tools have been developed, allowing editing at almost any desired location in the genome. Some examples include DNA base editors [73], epigenetic modifiers [74, 75], prime editors [76, 77], and transcription regulators [78, 79]. Fusion of various additional molecules with partially disrupted (nickase Cas9, nCas9) or nuclease-deficient (dead Cas9, dCas9) Cas9 has been used as a vehicle to deliver the CRISPR fusion protein to the target genomic site. RNA-targeting Cas proteins also enable a variety of RNA manipulations beyond simple RNA editing, such as RNA degradation, detection of ribonucleic acids and pathogens, single RNA base editing, and live imaging of RNA, which can be read in more detail in recently published reviews [72, 75]. Plants cope with stress through a range of finely tuned mechanisms, which involve both protein-coding genes and non-coding regions of the plant genome, along with various epigenetic mechanisms realized through the control of DNA packaging. The CRISPR-based tools described in this section can exploit the full range of molecular mechanisms mediated by these genomic elements.

4.2 Editing genes associated with transport and signaling of ABA and CKs

Major thriving areas of research include gene discovery (allele mining, investigation of cryptic genes) and introgression of new traits to achieve the desired goal—biotic/abiotic stress-resilient crops. Today, there is already a fairly large pool of works devoted to editing genes associated with the transport and signaling of ABA and CKs [75, 80]. Editing and transgenesis have helped to establish the functions of a number of genes associated with the ABA signaling pathway and their participation in the response to stress [81, 82]. For example, the enzymes SAPK1 and SAPK2 belonging to the SnRK2 family are members of the ABA signaling pathway in rice. Loss-of-function mutants of SAPK2 generated by CRISPR/Cas9 were insensitive to ABA [81]. The SAPK2 mutants displayed high sensitivity to dehydration and ROS, highlighting the role of SAPK2 in drought stress, the same as how CRISPR-edited OsERA1 mutant lines displayed enhanced tolerance to drought stress [83]. Another example is the work with histone acetyltransferase (HAT) enzyme that relaxes chromatin folding and promotes enhanced gene expression fused with dCas9 protein. Tools for gene activation and epigenetic modification combined with the CRISPR system made it possible to create the dCas9-HAT system, which increased the expression of AREB1 and as a result increased the resistance of Arabidopsis plants to drought [84]. CRISPR/Cas9 was successfully used to create new alleles of the OST2 gene in Arabidopsis, and as a result, edited plant lines carrying the new alleles exhibited an enhanced response to stress due to changes in stomatal closure under drought stress [85]. A number of genes have been shown to be involved in the negative regulation of plant responses to salinity and other abiotic stresses. Reducing the expression level of the RR22 gene, which encodes a type B response regulator (ARR B) involved in CK signaling, using the CRISPR/Cas9 system, made it possible to increase the tolerance of rice plants to soil salinity [86]. Additional examples of negative regulators research using editing tools include work in Arabidopsis and rice. Editing of the C/VIF1 gene encoding the fructosidase inhibitor protein 1 showed that it is a regulator of the response to ABA and is involved in the development of salt tolerance [87]. Editing of the RR9 and RR10 genes in rice, encoding proteins involved in the CK signaling pathway and associated with response regulators type A (ARR A), allowed to establish their function as negative regulators in response to salinity [88]. As recent work on AITR family genes has shown, targeting mutations in genes with redundant or unclear functions using CRISPR editing systems can help elucidate their role in plant stress biology [89, 90].

As it can be seen, various genome editing tools have been successfully used to study genes associated with plant stress resistance and to create stress-tolerant plants belonging not only to model plant species but also to plant species important for agriculture. The ever-expanding set of CRISPR tools allows you to make changes to any process occurring in a plant cell and thereby regulate the growth, development, and all life processes of plants, through precise and effective genetic engineering. Consistent changes and grouping of genes responsible for resistance to various types of stress, both biotic and abiotic, can help in the development of new lines for plant breeding. Accelerated identification of new genes, as well as the creation of gene-edited crops that do not fall under the regulatory requirements developed for transgenic plants, could be a step toward the next Green Transformation.

4.3 AREB/ABF and CKX gene families as potential targets for editing

CKX genes, which are key regulators of the level of CKs in plant cells and, accordingly, can influence the homeostasis of CKs in the cell, have long attracted the attention of researchers as providing ample opportunities for improving crops. Most studies investigating the function of CKX genes have been carried out using RNAi-based silencing or overexpression of CKX genes. Overexpression of AtCKX7 in the model plant results in shorter primary roots [91]. Overexpressing the AtCKS2 gene in oilseed *Brassica napus* increased the root-to-shoot ratio [92]. A number of studies have shown that reducing the expression level of CKX genes in some cases can lead to increased crop yields. For example, in barley, cotton, rice, and Arabidopsis, downregulation of CKX family genes through RNAi-based silencing or various genome editing systems, or with the help of mutations, has resulted in increased seed number and/or seed weight [93–96]. Also, in a number of works on editing genes of the OsCKX family in rice, it was shown that OsCKX genes serve as a link between CK and other plant hormones, in particular ABA [97, 98]. The perspectives of utilization of genome editing technologies to improve crop performance were discussed recently [80, 99]. The findings support the critical role of CKs in a variety of model plants.

There are significantly fewer studies on the AREB/ABF family. There is work to increase ABF2 expression using dCas9-HAT [84], but most of the research has been done on T-DNA-induced mutations in Arabidopsis obtained in the early 2000s [100–102]. In Arabidopsis, three members of the AREB/ABF family that respond to water stress and participate in the ABA signaling pathway, ABF2, ABF4 and ABF3, are the master transcription factors that co-regulate ABF-dependent ABA signaling and require ABA for full activation [100]. At the same time, the incomplete functional redundancy of ABF transcription factors gives reason to expect that differential manipulations of ABF can be used to create plants with the desired mode of ABA signaling, for example, to reduce trade-offs between ABA-induced stress tolerance and productivity.

Over the past few years, experimental evidence has been obtained on changes in DNA regions located at some distance from the site of T-DNA integration [103, 104]. This prompted a reconsideration of the relevance of using such mutations to identify the functions of genes of interest, since the manifestation of a mutation caused by the insertion of foreign DNA into the region of the gene under study and causing the loss of its function (knockout) can be masked by other insertions in regions remote from the region of the target gene. The development of new genome editing tools using CRISPR/Cas9 makes it possible to specifically make changes only in the target gene and obtain new series of knockouts for genes of interest. This work firstly examines the possibility of editing genes of the ABF family encoding the AREB1/ABF2, AREB2/ABF4, and ABF3 transcription factors using *Arabidopsis thaliana* as an example, taking into account the possible participation of other genes included in the network of regulation of abscisic acid biosynthesis. Secondly, the possibility of multiplex editing of CKX1 and CKX4 genes of *Arabidopsis thaliana* to establish their role in the response of plants to abiotic stress. Crossing the resulting mutants will make it possible to establish the details of the interaction between ABA and CKs.

5. Conclusion

The development and improvement of molecular biology methods by the beginning of the twenty first century stimulated the creation of modern tools that make it possible to modify plant genomes by targeted changes in the functioning of genes of interest. This opens up great opportunities for researchers to modify genes involved in the control of complex traits in plants, such as resistance to water deficiency. The use of genome editing to knockout individual genes that control plant response to various stress conditions, including water deficiency, will reveal the role of both regulatory genes encoding transcription factors for ABA biosynthesis and genes that provide interconnections between the signaling pathways of various phytohormones, in particular, the relationship between ABA and CKs.

Acknowledgements

This work was supported by the development program of National Research Tomsk State University (Priority-2030, project no. 2.1.2.22 "Genome editing as an innovative technology for studying the mechanisms of stress tolerance and increasing plant productivity under unfavorable environmental and climate changes").

Conflict of interest

The authors declare no conflict of interest.

Author details

Ilya Zlobin[1,2], Marina Efimova[1], Natalia Permykova[1,3], Irina Sokolova[3], Vladimir Kuznetsov[1,2] and Elena Deineko[1,3]*

1 National Research Tomsk State University, Tomsk, Russia

2 Institute of Plant Physiology, Russian Academy of Sciences, Moscow, Russia

3 Institute of Cytology and Genetics, Siberian Branch of the Russian Academy of Sciences, Novosibirsk, Russia

*Address all correspondence to: deineko@bionet.nsc.ru

IntechOpen

© 2023 The Author(s). Licensee IntechOpen. This chapter is distributed under the terms of the Creative Commons Attribution License (http://creativecommons.org/licenses/by/3.0), which permits unrestricted use, distribution, and reproduction in any medium, provided the original work is properly cited. (cc) BY

References

[1] Hammond WM, Adams HD. Dying on time: Traits influencing the dynamics of tree mortality risk from drought. Tree Physiology. 2019;**39**(6):906-909. DOI: 10.1093/treephys/tpz050

[2] Verslues PE, Bailey-Serres J, Brodersen C, Buckley TN, Conti L, Christmann A, et al. Burning questions for a warming and changing world: 15 unknowns in plant abiotic stress. The Plant Cell. 2023;**351**:67-108. DOI: 10.1093/plcell/koac263

[3] Agarwal PK, Gupta K, Lopato S, Agarwal P. Dehydration responsive element binding transcription factors and their applications for the engineering of stress tolerance. Journal of Experimental Botany. 2017;**68**(9):2135-2148. DOI: 10.1093/jxb/erx118

[4] Avni A, Golan Y, Shirron N, Shamai Y, Danin-Poleg Y, Gepstein S, et al. From survival to productivity mode: Cytokinins allow avoiding the avoidance strategy under stress conditions. Frontiers in Plant Science. 2020;**11**:879. DOI: 10.3389/fpls.2020.00879

[5] Hasan MM, Gong L, Nie Z-F, Li F-P, Ahammed GJ, Fang X-W. ABA-induced stomatal movements in vascular plants during dehydration and rehydration. Environmental and Experimental Botany. 2021;**186**:104436. DOI: 10.1016/j.envexpbot.2021.104436

[6] Brodribb T, Brodersen CR, Carriqui M, Tonet V, Dominguez CR, McAdam S. Linking xylem network failure with leaf tissue death. New Phytologist. 2021;**232**(1):68-79. DOI: 10.1111/nph.17577

[7] Brodribb TJ, McAdam SAM, Carins Murphy MR. Xylem and stomata, coordinated through time and space. Plant, Cell & Environment. 2017;**40**(6):872-880. DOI: 10.1111/pce.12817

[8] Martins SCV, McAdam SAM, Deans RM, DaMatta FM, Brodribb TJ. Stomatal dynamics are limited by leaf hydraulics in ferns and conifers: Results from simultaneous measurements of liquid and vapour fluxes in leaves. Plant, Cell & Environment. 2016;**39**(3):694-705. DOI: 10.1111/pce.12668

[9] Sussmilch FC, Schultz J, Hedrich R, Roelfsema MRG. Acquiring control: The evolution of stomatal signalling pathways. Trends in Plant Science. 2019;**24**(4):342-351. DOI: 10.1016/j.tplants.2019.01.002

[10] McAdam SAM, Brodribb TJ. Mesophyll cells are the main site of abscisic acid biosynthesis in water-stressed leaves. Plant Physiology. 2018;**177**(3):911-917. DOI: 10.1104/pp.17.01829

[11] Salvi P, Manna M, Kaur H, Thakur T, Gandass N, Bhatt D, et al. Phytohormone signaling and crosstalk in regulating drought stress response in plants. Plant Cell Reports. 2021;**40**:1305-1329. DOI: 10.1007/s00299-021-02683-8

[12] Sack L, John GP, Buckley TN. ABA accumulation in dehydrating leaves is associated with decline in cell volume, not turgor pressure. Plant Physiology. 2018;**176**(1):489-495. DOI: 10.1104/pp.17.01097

[13] Virlouvet L, Fromm M. Physiological and transcriptional memory in guard cells during repetitive dehydration stress. New Phytologist. 2015;**205**(2):596-607. DOI: 10.1111/nph.13080

[14] Sadhukhan A, Prasad SS, Mitra J, Siddiqui N, Sahoo L, Kobayashi Y, et al. How do plants remember drought? Planta. 2022;**256**(1):7. DOI: 10.1007/s00425-022-03924-0

[15] Buckley TN. How do stomata respond to water status? New Phytologist. 2019;**224**(1):21-36. DOI: 10.1111/nph.15899

[16] Flexas J, Carriquí M, Nadal M. Gas exchange and hydraulics during drought in crops: Who drives whom? Journal of Experimental Botany. 2018;**69**(16):3791-3795. DOI: 10.1093/jxb/ery235

[17] Polle A, Chen SL, Eckert C, Harfouche A. Engineering drought resistance in forest trees. Frontiers in Plant Science. 2019;**9**:1875. DOI: 10.3389/fpls.2018.01875

[18] Brodribb TJ, McAdam SAM. Abscisic acid mediates a divergence in the drought response of two conifers. Plant Physiology. 2013;**162**(3):1370-1377. DOI: 10.1104/pp.113.217877

[19] Brodribb TJ, McAdam SAM, Jordan GJ, Martins SCV. Conifer species adapt to low-rainfall climates by following one of two divergent pathways. National Academy of Sciences of the United States of America. 2014;**111**(40):14489-14493. DOI: 10.1073/pnas.140793011

[20] Heinrich S, Dippold MA, Werner C, Wiesenberg GLB, Kuzyakov Y, Glaser B. Allocation of freshly assimilated carbon into primary and secondary metabolites after in situ 13C pulse labelling of Norway spruce (*Picea abies*). Tree Physiology. 2015;**35**(11):1176-1191. DOI: 10.1093/treephys/tpv083

[21] Duursma RA, Blackman CJ, Lopéz R, Martin-StPaul K, Cochard H, Medlyn BE. On the minimum leaf conductance: Its role in models of plant water use, and ecological and environmental controls. New Phytologist. 2019;**221**(2):693-705. DOI: 10.1111/nph.15395

[22] Li X, Chen L, Forde BG, Davies WJ. The biphasic root growth response to abscisic acid in arabidopsis involves interaction with ethylene and auxin signalling pathways. Frontiers in Plant Science. 2017;**8**:1493. DOI: 10.3389/fpls.2017.01493

[23] Rowe JH, Topping JF, Liu J, Lindsey K. Abscisic acid regulates root growth under osmotic stress conditions via an interacting hormonal network with cytokinin, ethylene and auxin. New Phytologist. 2016;**211**(1):225-239. DOI: 10.1111/nph.13882

[24] Ranjan A, Sinha R, Singla-Pareek SL, Pareek A, Kumar SA. Shaping the root system architecture in plants for adaptation to drought stress. Physiologia Plantarum. 2022;**174**(2):e13651. DOI: 10.1111/ppl.13651

[25] Rosales MA, Maurel C, Nacry P. Abscisic acid coordinates dose-dependent developmental and hydraulic responses of roots to water deficit. Plant Physiology. 2019;**180**(4):2198-2211. DOI: 10.1104/pp.18.01546

[26] Calvo-Polanco M, Armada E, Zamarreño AM, García-Mina JM, Aroca R. Local root ABA/cytokinin status and aquaporins regulate poplar responses to mild drought stress independently of the ectomycorrhizal fungus Laccaria bicolor. Journal of Experimental Botany. 2019;**70**(21):6437-6446. DOI: 10.1093/jxb/erz389

[27] McAdam SAM, Brodribb TJ, Ross JJ. Shoot-derived abscisic acid promotes root growth. Plant, Cell & Environment. 2016;**39**(3):652-659. DOI: 10.1111/pce.12669

[28] Rosso L, Cantamessa S, Bergante S, Biselli C, Fricano A, Chiarabaglio PM. Responses to drought stress in poplar: What do we know and what can we learn? Life. 2023;**13**(2):533. DOI: 10.3390/life13020533

[29] Domec J-C, King JS, Carmichael MJ, Overby AT, Wortemann RR, Smith WK, et al. Root water gates and not changes in root structure provide new insights into plant physiological responses and adaptations to drought, flooding and salinity. bioRxiv. 27 Oct 2020. DOI: 10.1101/2020.10.27.357251

[30] De Diego N, Rodríguez JL, Dodd IC, Pérez-Alfocea F, Moncaleán P, Lacuesta M. Immunolocalization of IAA and ABA in roots and needles of radiata pine (*Pinus radiata*) during drought and rewatering. Tree Physiology. 2013;**33**(5):537-549. DOI: 10.1093/treephys/tpt033

[31] Brito C, Dinis L-T, Ferreira H, Moutinho-Pereira J, Correia CM. Foliar pre-treatment with abscisic acid enhances olive tree drought adaptability. Plants. 2020;**9**(3):341. DOI: 10.3390/plants9030341

[32] Gambetta GA, Knipfer T, Fricke W, McElrone AJ. Aquaporins and root water uptake. In: Chaumont F, Tyerman S, editors. Plant Aquaporins: From Transport to Signaling. Cham: Springer; 2017. pp. 133-153. DOI: 10.1007/978-3-319-49395-4_6

[33] Sánchez-Romera B, Ruiz-Lozano JM, Li G, Luu D-T, Martínez-Ballesta MC, Carvajal M. Enhancement of root hydraulic conductivity by methyl jasmonate and the role of calcium and abscisic acid in this process. Plant, Cell & Environment. 2014;**37**(4):995-1008. DOI: 10.1111/pce.12214

[34] Parent B, Hachez C, Redondo E, Simonneau T, Chaumont F, Tardieu F. Drought and abscisic acid effects on aquaporin content translate into changes in hydraulic conductivity and leaf growth rate: A trans-scale approach. Plant Physiology. 2009;**149**(4):2000-2012. DOI: 10.1104/pp.108.130682

[35] Prado K, Maurel C. Regulation of leaf hydraulics: From molecular to whole plant levels. Frontiers in Plant Science. 2013;**4**:255. DOI: 10.3389/fpls.2013.00255

[36] Aloni R. The role of hormones in controlling vascular differentiation. In: Fromm J, editor. Cellular Aspects of Wood Formation. Berlin, Heidelberg: Springer; 2013. pp. 99-139. DOI: 10.1007/978-3-642-36491-4_4

[37] Buttò V, Deslauriers A, Rossi S, Rozenberg P, Shishov V, Morin H. The role of plant hormones in tree-ring formation. Trees. 2020;**34**:315-335. DOI: 10.1007/s00468-019-01940-4

[38] Mwange KNK, Hou H-W, Wang Y-Q, He X-Q, Cui K-M. Opposite patterns in the annual distribution and time-course of endogenous abscisic acid and indole-3-acetic acid in relation to the periodicity of cambial activity in Eucommia ulmoides Oliv. Journal of Experimental Botany. 2005;**56**(413):1017-1028. DOI: 10.1093/jxb/eri095

[39] Rodriguez-Zaccaro FD, Groover A. Wood and water: How trees modify wood development to cope with drought. Plants, People, Planet. 2019;**1**(4):346-355. DOI: 10.1002/ppp3.29

[40] Blackman CJ, Brodribb TJ, Jordan GJ. Leaf hydraulics and drought stress: Response, recovery and survivorship in four woody temperate plant species. Plant, Cell & Environment. 2009;**32**(11):1584-1595. DOI: 10.1111/j.1365-3040.2009.02023.x

[41] Lovisolo C, Perrone I, Hartung W, Schubert A. An abscisic acid-related

reduced transpiration promotes gradual embolism repair when grapevines are rehydrated after drought. New Phytologist. 2008;**180**(3):642-651. DOI: 10.1111/j.1469-8137.2008.02592.x

[42] Tombesi S, Nardini A, Frioni T, Soccolini M, Zadra C, Farinelli D. Stomatal closure is induced by hydraulic signals and maintained by ABA in drought-stressed grapevine. Scientific Reports. 2015;**5**(1):12449. DOI: 10.1038/srep12449

[43] Kambona CM, Koua PA, Léon J, Ballvora A. Stress memory and its regulation in plants experiencing recurrent drought conditions. Theoretical and Applied Genetics. 2023;**136**(2):26. DOI: 10.1007/s00122-023-04313-1

[44] Zhang H, Zhao Y, Zhu JK. Thriving under stress: How plants balance growth and the stress response. Developmental Cell. 2020;**55**(5):529-543. DOI: 10.1016/j.devcel.2020.10.012

[45] Querejeta JI, Ren W, Prieto I. Vertical decoupling of soil nutrients and water under climate warming reduces plant cumulative nutrient uptake, water-use efficiency and productivity. New Phytologist. 2021;**230**(4):1378-1393. DOI: 10.1111/nph.17258

[46] Waadt R, Seller CA, Hsu P-K, Takahashi Y, Munemasa S, Schroeder JI. Plant hormone regulation of abiotic stress responses. Nature Reviews Molecular Cell Biology. 2022;**23**(10):680-694. DOI: 10.1038/s41580-022-00479-6

[47] Hai NN, Chuong NN, Tu NHC, Kisiala A, Hoang XLT, Thao NP. Role and regulation of cytokinins in plant response to drought stress. Plants. 2020;**9**(4):422. DOI: 10.3390/plants9040422

[48] Prerostova S, Dobrev PI, Gaudinova A, Knirsch V, Körber N, Pieruschka R. Cytokinins: Their impact on molecular and growth responses to drought stress and recovery in arabidopsis. Frontiers in Plant Science. 2018;**9**:655. DOI: 10.3389/fpls.2018.00655

[49] Romanov GA. Perception, transduction and crosstalk of auxin and cytokinin signals. International Journal of Molecular Sciences. 2022;**23**(21):13150. DOI: 10.3390/ijms232113150

[50] Cortleven A, Leuendorf JE, Frank M, Pezzetta D, Bolt S, Schmülling T. Cytokinin action in response to abiotic and biotic stresses in plants. Plant, Cell & Environment. 2019;**42**(3):998-1018. DOI: 10.1111/pce.13494

[51] De Diego N, Pérez-Alfocea F, Cantero E, Lacuesta M, Moncaleán P. Physiological response to drought in radiata pine: Phytohormone implication at leaf level. Tree Physiology. 2012;**32**(4):435-449. DOI: 10.1093/treephys/tps029

[52] Immanen J, Nieminen K, Smolander O-P, Kojima M, Serra JA, Koskinen P. Cytokinin and auxin display distinct but interconnected distribution and signaling profiles to stimulate cambial activity. Current Biology. 2016;**26**(15):1990-1997

[53] Hu L, Wang Z, Huang B. Effects of cytokinin and potassium on stomatal and photosynthetic recovery of Kentucky bluegrass from drought stress. Crop Science. 2013;**53**(1):221-231. DOI: 10.2135/cropsci2012.05.0284

[54] Qin H, He L, Huang R. The coordination of ethylene and other hormones in primary root development. Frontiers in Plant Science. 2019;**10**:874. DOI: 10.3389/fpls.2019.00874

[55] Waidmann S, Sarkel E, Kleine-Vehn J. Same, but different: Growth responses of primary and lateral roots. Journal of Experimental Botany. 2020;**71**(8):2397-2411. DOI: 10.1093/jxb/eraa027

[56] Kurepa J, Smalle JA. Auxin/cytokinin antagonistic control of the shoot/root growth ratio and its relevance for adaptation to drought and nutrient deficiency stresses. International Journal of Molecular Sciences. 2022;**23**(4):1933. DOI: 10.3390/ijms23041933

[57] Ramireddy E, Nelissen H, Leuendorf JE, Lijsebettens MV, Inzé D, Schmülling T. Root engineering in maize by increasing cytokinin degradation causes enhanced root growth and leaf mineral enrichment. Plant Molecular Biology. 2021;**106**:555-567. DOI: 10.1007/s11103-021-01173-5

[58] Gelvin SB. Integration of agrobacterium T-DNA into the plant genome. Annual Review of Genetics. 2017;**51**:195-217. DOI: 10.1146/annurev-genet-120215-035320

[59] Lee J, Chin JH, Ahn SN, Koh H. Brief history and perspectives on plant breeding. In: Current Technologies in Plant Molecular Breeding. Dordrecht: Springer; 2015. 343 p. DOI: 10.1007/978-94-017-9996-6

[60] Mahfouz MM, Li L. TALE nucleases and next generation GM crops. GM Crops. 2011;**2**(2):99-103. DOI: 10.4161/gmcr.2.2.17254

[61] Wood AJ, Lo T-W, Zeitler B, Pickle CS, Ralston EJ, Lee AH, et al. Targeted genome editing across species using ZFNs and TALENs. Science. 2011;**333**:307. DOI: 10.1126/science.1207773

[62] Shukla VK, Doyon Y, Miller JC, Dekelver RC, Moehle EA, Worden SE, et al. Precise genome modification in the crop species *Zea mays* using zinc-finger nucleases. Nature. 2009;**459**:437-441. DOI: 10.1038/nature07992

[63] Townsend JA, Wright DA, Winfrey RJ, Fu F, Maeder ML, Joung JK, et al. High-frequency modification of plant genes using engineered zinc-finger nucleases. Nature. 2009;**459**:442-445. DOI: 10.1038/nature07845

[64] Osakabe K, Osakabe Y, Toki S. Site-directed mutagenesis in arabidopsis using custom-designed zinc finger nucleases. Proceedings of the National Academy of Sciences of the United States of America. 2010;**107**:12034-12039. DOI: 10.1073/pnas.1000234107

[65] Carlson DF, Tan W, Lillico SG, Stverakova D, Proudfoot C, Christian M, et al. Efficient TALEN-mediated gene knockout in livestock. Proceedings of the National Academy of Sciences of the United States of America. 2012;**109**:17382-17387. DOI: 10.1073/pnas.1211446109

[66] Zhang Y, Zhang F, Li X, Baller JA, Qi Y, Starker CG, et al. Transcription activator-like effector nucleases enable efficient plant genome engineering. Plant Physiology. 2013;**161**:20-27. DOI: 10.1104/pp.112.205179

[67] Li J-F, Norville JE, Aach J, McCormack M, Zhang D, Bush J, et al. Multiplex and homologous recombination–mediated genome editing in arabidopsis and *Nicotiana benthamiana* using guide RNA and Cas9. Nature Biotechnology. 2013;**31**:688-691. DOI: 10.1038/nbt.2654

[68] Jinek M, Chylinski K, Fonfara I, Hauer M, Doudna JA, Charpentier E. A. A programmable dual-RNA—Guided DNA endonuclease in adaptive bacterial immunity. Science 2012;337:816-822. DOI: 10.1126/science.1225829

[69] Cong L, Ran FA, Cox D, Lin S, Barretto R, Habib N, et al. Multiplex genome engineering using CRISPR/Cas systems. Science. 2013;**339**:819-823. DOI: 10.1126/science.1231143

[70] Mali P, Yang L, Esvelt KM, Aach J, Guell M, DiCarlo JE, et al. RNA-guided human genome engineering via Cas9. Science. 2013;**339**:823-826. DOI: 10.1126/science.1232033

[71] Li W, Teng F, Li T, Zhou Q. Simultaneous generation and germline transmission of multiple gene mutations in rat using CRISPR-Cas systems. Nature Biotechnology. 2013;**31**:684-686. DOI: 10.1038/nbt.2652

[72] Pramanik D, Shelake RM, Kim MJ, Kim JY. CRISPR-mediated engineering across the central dogma in plant biology for basic research and crop improvement. Molecular Plant. 2021;**14**:127-150. DOI: 10.1016/j.molp.2020.11.002

[73] Jeong YK, Song B, Bae S. Current status and challenges of DNA base editing tools. Molecular Therapy. 2020;**28**:1938-1952. DOI: 10.1016/j.ymthe.2020.07.021

[74] Shelake RM, Pramanik D, Kim JY. Evolution of plant mutagenesis tools: A shifting paradigm from random to targeted genome editing. Plant Biotechnology Reports. 2019;**13**:423-445. DOI: 10.1007/s11816-019-00562-z

[75] Shelake RM, Kadam US, Kumar R, Pramanik D, Singh AK, Kim JY. Engineering drought and salinity tolerance traits in crops through CRISPR-mediated genome editing: Targets, tools, challenges, and perspectives. Plant Communications. 2022;**3**:100417. DOI: 10.1016/j.xplc.2022.100417

[76] Huang TK, Puchta H. Novel CRISPR/Cas applications in plants: From prime editing to chromosome engineering. Transgenic Research. 2021;**30**:529-549. DOI: 10.1007/s11248-021-00238-x

[77] Lin Q, Jin S, Zong Y, Yu H, Zhu Z, Liu G, et al. High-efficiency prime editing with optimized, paired pegRNAs in plants. Nature Biotechnology. 2021;**39**(8):923-927. DOI: 10.1038/s41587-021-00868-w

[78] Piatek A, Ali Z, Baazim H, Li L, Abulfaraj A, Al-Shareef S, et al. RNA-guided transcriptional regulation in planta via synthetic dCas9-based transcription factors. Plant Biotechnology Journal. 2015;**13**:578-589. DOI: 10.1111/pbi.12284

[79] Rehman RS, Zafar SA, Ali M, Pasha AN, Naveed MS, Waseem M, et al. CRISPR-Cas mediated genome editing: A paradigm shift towards sustainable agriculture and biotechnology. Asian Plant Research Journal. 2022;**9**:27-49. DOI: 10.9734/aprj/2022/v9i130197

[80] Mandal S, Ghorai M, Anand U, Roy D, Kant N, Mishra T, et al. Cytokinins: A genetic target for increasing yield potential in the CRISPR era. Frontiers in Genetics. 2022;**13**:1-12. DOI: 10.3389/fgene.2022.883930

[81] Lou D, Wang H, Liang G, Yu D. OsSAPK2 confers abscisic acid sensitivity and tolerance to drought stress in rice. Frontiers in Plant Science. 2017;**8**:1-15. DOI: 10.3389/fpls.2017.00993

[82] He QY, Jin JF, Lou HQ, Dang FF, Xu JM, Zheng SJ, et al. Abscisic acid-dependent PMT1 expression regulates salt tolerance by alleviating abscisic acid-mediated reactive oxygen species production in arabidopsis. Journal of Integrative Plant Biology. 2022;**64**:1803-1820. DOI: 10.1111/jipb.13326

[83] Ogata T, Ishizaki T, Fujita M, Fujita Y. CRISPR/Cas9-targeted mutagenesis of

OsERA1 confers enhanced responses to abscisic acid and drought stress and increased primary root growth under nonstressed conditions in rice. PLoS One. 2020;**15**:1-12. DOI: 10.1371/journal.pone.0243376

[84] Roca Paixão JF, Gillet FX, Ribeiro TP, Bournaud C, Lourenço-Tessutti IT, Noriega DD, et al. Improved drought stress tolerance in arabidopsis by CRISPR/dCas9 fusion with a histone acetyltransferase. Scientific Reports. 2019;**9**:1-9. DOI: 10.1038/s41598-019-44571-y

[85] Osakabe Y, Watanabe T, Sugano SS, Ueta R, Ishihara R, Shinozaki K, et al. Optimization of CRISPR/Cas9 genome editing to modify abiotic stress responses in plants. Scientific Reports. 2016;**6**:1-10. DOI: 10.1038/srep26685

[86] Zhang A, Liu Y, Wang F, Li T, Chen Z, Kong D, et al. Enhanced rice salinity tolerance via CRISPR/Cas9-targeted mutagenesis of the OsRR22 gene. Molecular Breeding. 2019;**39**:47. DOI: 10.1007/s11032-019-0954-y

[87] Yang W, Chen S, Cheng Y, Zhang N, Ma Y, Wang W, et al. Cell wall/vacuolar inhibitor of fructosidase 1 regulates ABA response and salt tolerance in arabidopsis. Plant Signaling & Behavior. 2020;**15**:4. DOI: 10.1080/15592324.2020.1744293

[88] Wang WC, Lin TC, Kieber J, Tsai YC. Response regulators 9 and 10 negatively regulate salinity tolerance in Rice. Plant & Cell Physiology. 2019;**60**:2549-2563. DOI: 10.1093/pcp/pcz149

[89] Chen S, Zhang N, Zhou G, Hussain S, Ahmed S, Tian H, et al. Knockout of the entire family of AITR genes in Arabidopsis leads to enhanced drought and salinity tolerance without fitness costs. BMC Plant Biology. 2021;**21**:1-15. DOI: 10.1186/s12870-021-02907-9

[90] Wang T, Xun H, Wang W, Ding X, Tian H, Hussain S, et al. Mutation of GmAITR genes by CRISPR/Cas9 genome editing results in enhanced salinity stress tolerance in soybean. Frontiers in Plant Science. 2021;**12**:1-15. DOI: 10.3389/fpls.2021.779598

[91] Köllmer I, Novák O, Strnad M, Schmülling T, Werner T. Overexpression of the cytosolic cytokinin oxidase/dehydrogenase (CKX7) from Arabidopsis causes specific changes in root growth and xylem differentiation. The Plant Journal. 2014;**78**:359-371. DOI: 10.1111/tpj.12477

[92] Nehnevajova E, Ramireddy E, Stolz A, Gerdemann-Knörck M, Novák O, Strnad M, et al. Root enhancement in cytokinin-deficient oilseed rape causes leaf mineral enrichment, increases the chlorophyll concentration under nutrient limitation and enhances the phytoremediation capacity. BMC Plant Biology. 2019;**19**:1-15. DOI: 10.1186/s12870-019-1657-6

[93] Zalewski W, Galuszka P, Gasparis S, Orczyk W, Nadolska-Orczyk A. Silencing of the HvCKX1 gene decreases the cytokinin oxidase/dehydrogenase level in barley and leads to higher plant productivity. Journal of Experimental Botany. 2010;**61**:1839-1851. DOI: 10.1093/jxb/erq052

[94] Bartrina I, Otto E, Strnad M, Werner T, Schmülling T. Cytokinin regulates the activity of reproductive meristems, flower organ size, ovule formation, and thus seed yield in *Arabidopsis thaliana*. The Plant Cell. 2011;**23**:69-80. DOI: 10.1105/tpc.110.079079

[95] Li S, Zhao B, Yuan D, Duan M, Qian Q, Tang L, et al. Rice zinc finger protein DST enhances grain production through controlling Gn1a/OsCKX2 expression.

Proceedings of the National Academy of Sciences of the United States of America. 2013;**110**:3167-3172. DOI: 10.1073/pnas.1300359110

[96] Zhao J, Bai W, Zeng Q, Song S, Zhang M, Li X, et al. The abscisic acid–responsive element binding factors MAPKKK18 module regulates abscisic acid–induced leaf senescence in arabidopsis. The Journal of Biological Chemistry. 2023;**299**:103060. DOI: 10.1016/j.jbc.2023.103060

[97] Tao Y, Miao J, Wang J, Li W, Xu Y, Wang F, et al. RGG1, involved in the cytokinin regulatory pathway, controls grain size in rice. Rice. 2020;**13**:76. DOI: 10.1186/s12284-020-00436-x

[98] Zhang W, Peng K, Cui F, Wang D, Zhao J, Zhang Y, et al. Cytokinin oxidase/dehydrogenase OsCKX11 coordinates source and sink relationship in rice by simultaneous regulation of leaf senescence and grain number. Plant Biotechnology Journal. 2021;**19**:335-350. DOI: 10.1111/pbi.13467

[99] Mahto RK, Singh C, Chandana BS, Singh RK, Verma S, Gahlaut V. Chickpea biofortification for cytokinin dehydrogenase via genome editing to enhance abiotic-biotic stress tolerance and food security. Frontiers in Genetics. 2022;**13**:900324. DOI: 10.3389/fgene.2022.900324

[100] Yoshida T, Fujita Y, Sayama H, Kidokoro S, Maruyama K, Mizoi J, et al. AREB1, AREB2, and ABF3 are master transcription factors that cooperatively regulate ABRE-dependent ABA signaling involved in drought stress tolerance and require ABA for full activation. The Plant Journal. 2010;**61**:672-685. DOI: 10.1111/j.1365-313X.2009.04092.x

[101] Du J, Zhu X, He K, Kui M, Zhang J, Han X, et al. CONSTANS interacts with and antagonizes ABF transcription factors during salt stress under long-day conditions. Plant Physiology. 2023;**193**:1675-1694. DOI: 10.1093/plphys/kiad370

[102] Zhao J, Bai W, Zeng Q, Song S, Zhang M, Li X, et al. Moderately enhancing cytokinin level by down-regulation of GhCKX expression in cotton concurrently increases fiber and seed yield. Molecular Breeding. 2015;**35**:60. DOI: 10.1007/s11032-015-0232-6

[103] Pucker B, Kleinbölting N, Weisshaar B. Large scale genomic rearrangements in selected Arabidopsis thaliana T-DNA lines are caused by T-DNA insertion mutagenesis. BMC Genomics. 2021;**22**:1-21. DOI: 10.1186/s12864-021-07877-8

[104] Edwards B, Hornstein ED, Wilson NJ, Sederoff H. High-throughput detection of T-DNA insertion sites for multiple transgenes in complex genomes. BMC Genomics. 2022;**23**:1-20. DOI: 10.1186/s12864-022-08918-6

Chapter 2

Organic Farming to Mitigate Abiotic Stresses under Climate Change Scenario

Saikat Biswas, Rupa Das and Lay Lay Nwe

Abstract

Climate change is resultant from modern-day chemical agriculture, which is creating negative impacts on crop production. Global agriculture is now facing various problems arising due to abiotic stresses such as flood, drought, temperature extremes, light extremes, salinity, heavy metal stress, nutrient toxicity/deficiency. These stresses not only hamper the growth and production but also reduce the quality of crops through morphological, physiological, biochemical changes and synthesis of ROS. Further, they negatively impact on entire environment specially soil health. Deterioration of yield and quality often occurs due to lack of essential inputs to plants under abiotic stresses. Although plants adopt defensive mechanisms, such abiotic stresses need to be addressed properly with various eco-friendly organic farming approaches. Different organic inputs like organic manures, biofertilizers, bio-priming with microorganisms, bio-stimulants (seaweed extracts, humic acid, micro-organisms, etc.), mulches, biochar are known to alleviate abiotic stresses under climate change scenario. Further, various organic agronomic practices viz. crop rotation, intercropping, tillage, sowing methods and time, nutrient, water and intercultural operations, use of PGPB, organic formulations, grafting, selection of resistant/tolerant varieties and other scientific/wise uses of organic inputs can mitigate/escape the negative impacts of abiotic stresses resulting in upliftment in crop production as well as the quality of produce.

Keywords: abiotic stresses, agronomic management, climate change, crop growth, organic farming practices, production

1. Introduction

Food scarcity is a major challenge in today's agriculture. In order to meet the food demand of ever-increasing population, worldwide, farmers are aiming to improve agricultural productivity at the expanse of environment through application of chemical fertilizers and pesticides. Unscientific and over use of chemicals and other management practices degrades soil, water and other valuable natural resources leading to climate change scenario which is a great concern for sustainable agriculture. Further, shrinkage of agricultural land due to population growth, aim/migration for alternative

IntechOpen

job, urbanization, deforestation, anti-environmental anthropogenic activities, etc. are creating major issues of agriculture and urging for improvement of agricultural productivity and fulfillment of this urge is questionable under climate change scenario as plant has sessile growth habit. Crop growth mostly depends on interaction between genetic trait of variety with growing environment. Climate change, therefore, exerts various stresses on crops and affects crop negatively. The stresses can be biotic (living) and abiotic (non-living) stresses which often put both sole and combined impact on crop.

Abiotic stresses such as drought, flood, salinity, temperature extremes (hot/cold), heavy metals, light, wind, nutrients/chemicals, etc. due to climate change decide distribution of plants in various environmental conditions [1] and thereafter, affect crop growth especially at reproductive stage resulting in poor crop productivity throughout the world [2]. Abiotic stresses also trigger various biotic stresses leading to poor crop productivity through disruption of seed germination, vegetative growth, dry matter production and its translocation to reproductive parts [3]. World experiences around 70% yield loss due to abiotic stresses [4]. Severity in abiotic stresses causes imbalance between demand and supply of nutrients, inactivation of enzymatic activities, suppression of various genes responsible for the quality expression, etc. [5] resulting loss of yield and quality of crop through hampering crop from morphological to molecular levels [6].

These abiotic stresses indeed are serious barriers in front of food security of global population and therefore, suitable strategies are highly needed to cope with these and to achieve good crop growth, yield and quality under climate change scenario. Although few mechanisms like escaping stress, stress avoidance and stress tolerance are done by plants through making various molecular, cellular and physiological changes, there is need to explore and adopt various strategies like traditional and modern breeding approaches, agronomic management practices, exogenous applications of stress tolerating compounds, etc. to mitigate harmful impacts of abiotic stresses on crop to a high extent. Agronomic management strategies cover various technologies including organic farming approaches to alleviate abiotic stresses. Organic farming consists of chemical excluded farming practices which mostly rely on natural and organic inputs/products to improve crop growth and yield as well as other allied sectors. Various organic farming inputs (manures, biofertilizers, crop residues, bio-stimulants, etc.) and practices (selection of varieties, tillage, sowing, nutrient, weed, water management practices, etc.) play a key role in addressing various abiotic stresses and allows the crop to grow well by coping up the climate change situation. Although the published information is less in this regard, an insight knowledge on organic farming activates against abiotic stresses is highly needed. Therefore, an attempt was made in this chapter to highlight negative impacts of abiotic stresses on crop and their mitigation strategies through various organic farming approaches.

2. Various abiotic stresses

Abiotic stress is resulted due to the negative influence of physical or chemical environment on biological organisms either alone or in various forms of interaction. In agriculture, crop production is highly hampered due to abiotic stresses. They individually or as combination impair the normal metabolisms and other physiological functions in plants and thereby, affect crop growth and development. Combined influence of various stresses is more pronounced on crop production than their individual adverse effect. There are different types of stresses viz. drought, flood, heat and cold stresses, heavy metals, toxicity due to nutrients and pesticides, high light, low light, UV exposure, photo-inhibition, shade, wind velocity, air pollution, etc. which negatively impact on crop plants (**Figure 1**). In the following section, these stresses are briefly highlighted.

2.1 Drought stress

Among the different natural resources, water is now highly precious and scarce for living organisms including crop. Water regulates various physiological and biochemical activities of plant like photosynthesis, transpiration, nutrient uptakes, translocation of assimilates, etc. Plant growth, internal activities are seriously hampered if water availability deviates from normal which is now a common phenomenon throughout the globe as an effect of climate change due to unscientific, non-eco-friendly anthropogenic activities. Water stress mostly occurs in the form drought and flood. Drought or water scarcity may arise due to various reasons such as long period of no occurrence or less intensity of rainfall from usual, low river and stream flows, reduced ground water table, etc. in a region. In agriculture, during crop growth stages, drought may arise due to late onset and early cessation of rainfall, break of monsoon for long period, less availability of irrigation water, faulty or no water conservation practices/structures resulting in serious damage to crop growth and yield. Moreover, Under the situation of soil moisture availability, if salt concentration is high in soil, plant can't uptake water from soil properly and even, exosmosis occurs. This situation, thus creates apparent drought. High temperature triggers evapotranspiration as a part of internal cooling process, resulting in drought or water deficiency. Further, drought can also be resulted from low temperature, under which water freezes in the intercellular spaces creating protoplasmic dehydration and death of cell and eventually, the plant. Altogether, drought affects plant's germination and normal functioning.

2.2 Flood stress

When water availability becomes unnecessarily high as compared to normal for a particular period in an area, flood occurs. It may be resulted from sudden outburst of cloud coupled with excessive rainfall for a short time period (flash flood) or due to

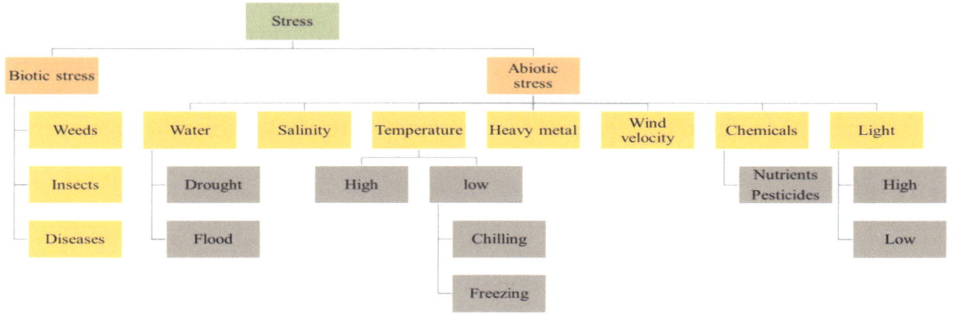

Figure 1.
Classification of various biotic and abiotic stresses [7].

continuous rainfall for few days or high water-table or overflow of river, ponds and dams associated with less drainage facility. Flash flood lasts for a very less time period from a day to only few weeks. However, deep water flood lasts for a longer period of time.

2.3 Salinity stress

Throughout the globe specially during arid and semi-arid areas, salinity is a major issue. It arises in areas where potential evapotranspiration is greater than the rainfall as well as insufficient leaching of salts beyond rhizospheric zone owing from poor rainfall. Presence of excess salts in soil drastically hampers the crop growth [8]. Soil salinity can be developed by both natural phenomena (Weathering of rocks, flooding and intrusion of sea water to agricultural land, seepage of saline water, wind blow, etc.) and human induced activities (poor water quality of irrigation, deforestation, overgrazing, intensive cropping, etc.). Salinity is indicated by electrical conductivity (EC). Usually, soil having EC > 4 dS/m, exchangeable sodium percentage (ESP) < 15.0 and pH <8.5 is called as saline soil [9]. Saline soil contains chloride, sulfate salts of sodium, magnesium and calcium ions. Presence of these salts in excessive quantities deteriorates soil health through changing cation exchange capacity, negatively impacting soil micro-organisms' survival, multiplication and activities, disrupting soil physical properties through deflocculation and reduction of hydraulic conductivity, etc.

2.4 Temperature stress

Temperature stress indicates both rise and fall of temperature from normal. Sudden change in temperature occurs due to climate change. Specifically, heat stress or high temperature situation arise due to due to global warming and anthropogenic activities resulting in change biodiversity, crop ecosystem, impairment of crop growth and production especially in areas of tropics and sub-tropics. Heat stress results in respiration greater than photosynthesis causing starvation injury through deficit of food reserves in plants. According to different degree of high temperature tolerance, plants are categorized as psychrophiles (up to 15–20°C), mesophiles (up to 35–45°C) and thermophiles (up to 45–100°C) [10].

In contrast to heat stress, an opposite phenomenon known as cold stress or low temperature occurs mostly in temperate areas. There are two types of cold stress viz. chilling stress and freezing stress both affecting the crop's physiological, biochemical activities and eventually, hampering crop's growth, yield and quality. Similar to heat stress, plants are also grouped into three based on cold stress tolerance: Chilling sensitive (Plants are extremely sensitive above 0°C and below 15°C), chilling resistant (plants can tolerate low temperature but highly suffer under formation of ice crystals in intra and inter cellular spaces) and frost resistant (plants are tolerant to extremely low temperature).

2.5 Heavy metal toxicity

Heavy metals impart mutagenic effects on plants by contaminating irrigation water, food chain and environment [11]. These are inorganic, non-biodegradable compounds with atomic mass >20 and density >5 g/cm^3. The source of heavy metals in the soil is use of irrigation water from contaminated area, excessive application of chemical fertilizer and pesticides. Plants absorb heavy metals from soil through roots. Ag, Cr, Cd, As, Sb, Pb, Se, and Hg are some major heavy metals which at high concentrations are non-essential and thereby, hamper soil quality and plant's normal functioning. Other than these, there are some essential elements viz. Zn, Cu, Ni, Fe, Co, etc. which at high concentrations create heavy metal toxicity in soil and plant.

2.6 Light stress

Light is essential resource not only for plant growth but also for all life. In fact, harnessing of high amount of solar energy is the prime aim of crop production. However, excessive or low light can cause negative impact on crop such as poor crop growth, wilting, dwarfing, less photosynthesis, cell damage, low productivity and quality and even death of the plant.

2.7 Wind velocity

Wind plays a major role in maintenance of aeration, pollination, etc. in crop's microclimate. However, high wind velocity over the cropped area can exert stress on crop. Wind velocity occurs due to movement of wind from one direction to other at a particular speed. It can create high evapo-transpiration, sand injury, crop lodging, pollen shedding, loss of pollen through desiccation, etc.

2.8 Chemical toxicity

Continuous dependence on chemical based inorganic fertilizers and pesticides specially after green revolution is a great concern now in present day intensive agriculture condition. Further, rapid industrialization and excessive use of untreated sewage water hampers crop's growth and productivity through exerting detrimental impacts of the chemical toxicity on the soil- plant-atmospheric continuum.

2.9 Nutrient toxicity/deficiency

Nutrient toxicity or deficiency resulted from excessive or scarce application of fertilizers and manures as well as soil own nutritional status impairs plant growth, productivity and quality of the crop. This situation is very common in today's intensive agriculture due to non-judicious, unscientific nutrient management practices by the farmers.

3. Negative impacts of abiotic stresses on crop

Plants are negatively impacted by abiotic stress. In most cases, abiotic stresses exert combined impact on crop plants and it causes more harm over individual impact of stress. Hydrogen peroxide, hydroxyl radicals, superoxide radicals, singlet oxygen and other reactive oxygen species (ROS) are synthesized under various abiotic stresses, specially under drought stress. In combinations, these cause lipid peroxidation, protein oxidation etc. and affect nucleic acids and enzyme activity resulting in death of cell. Accordingly, plants adopt defensive mechanisms against stresses. For instance, under drought stress, partial or complete closure of stomata is the one such adoptive approach by plants, which further restricts entry of sunlight, CO_2 and impairs electron flow through electron transport chain, resulting in decline in photosynthesis. Various negative impacts of abiotic stress on plants are shown in **Figure 2** and **Table 1** and highlighted hereunder.

3.1 Drought stress

Drought stress arises under water scarcity. It hampers seed germination as well as early stand establishment of a crop arising through depletion of seed reserves and mechanical obstruction by the hard soil under drought, resulting in poor vegetative growth and yield of crop. The various impacts of drought on physiological and biochemical activities of plants are shown in **Figure 3**. When drought arises, cell solutes concentrations increase due to less water uptake and it not only causes high intra- and inter-competitions for water among crop plants and between crop and weeds, but also exerts toxicity on plants. Further, under drought condition, nutrients show variations in their availability for plant's uptake. Few nutrients become more available (viz. nitrogen) and few become unavailable or less available (viz. phosphorus), while no distinct impact of drought occurs on some nutrients (viz. potassium). This creates alterations in nutrient uptake by plants resulting in impairment of nutrient metabolisms in cell [36]. Under drought stress, as the activities of enzymes such as nitrate reductase, glutamine synthetase, etc. decrease, ammonia assimilation to organic form is restricted. Among the different categories of plants, C_4 ones suffer more than C_3 plants due to closure of stomata resulting in less photosynthesis [37].

3.2 Flood stress

When flood occurs, anaerobic situation arises due to water logging or submergence, which further causes depletion in oxygen as well as restriction of movement of oxygen and other gases in root zone of plant. As a consequence, chlorosis of plant leaves and decay/death of cell occur. Less root respiration, poor root proliferation and

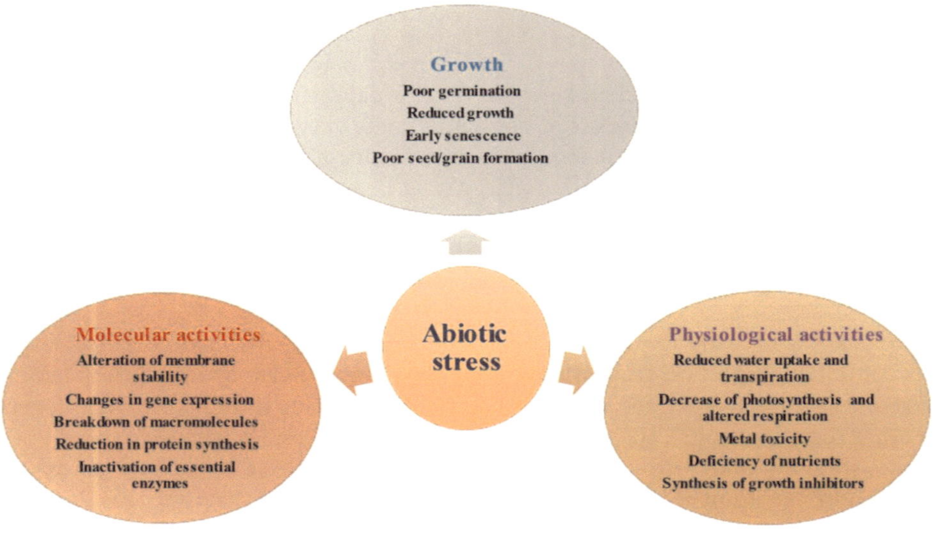

Figure 2.
Influence of abiotic stress on plant [7].

Crop	Abiotic stress	References	Crop	Abiotic stress	References
Lentil	Drought stress	[12]	Lentil	Heat stress	[13]
Chick pea		[14]	Wheat		[15–18]
Soybean		[19]	Rice		[20]
Common bean		[21]	Ground nut		[13]
Mung bean		[22]	Chick pea		[13]
Faba bean		[23]	Pea		[24]
Barley		[5]	Pigeon pea		[13]
Wheat		[25]	Cow pea		[13]
Cotton		[26]	Soybean		[13]
Maize		[27]	Mung bean		[24]
Spotted bean		[12]	Common bean		[28]
Black gram		[29]	Broad bean		[24]
Cow pea		[30]	Lupin		[24]
Pigeon pea		[23]	Groundnut, chickpea, green gram, soybean, pigeon pea	Cd stress	[31]
Lupin		[23]	Grass pea, chick pea	Pb stress	[31]
Soybean	Salinity stress	[31]	Chick pea, green gram	Cr stress	[31]
Chick pea		[31]	Pea, lentil, soybean, black gram	Hg stress	[31]
Lentil		[31]	Pea, chick pea, cowpea, green gram	Cu stress	[31]
Mung bean		[32]	Chick pea, cowpea, pigeon pea	Ni stress	[31]
Faba bean		[33]	Cowpea, chick pea	Zn stress	[31]
Wheat		[34]	Pea, chick pea	As stress	[31]
Soybean	Cold stress	[28]			
Rice		[35]			
Broad bean, Pea		[28]			
Chick pea		[28]			

Table 1.
Negative impacts of abiotic stresses on production of various crops.

other physiological disorders are some common phenomena visible under flood condition. Various negative impacts of flood are shown in **Figure 3**.

3.3 Salinity stress

Salinity creates two prime impacts on plants viz. osmotic stress and ion toxicity. Under the situation of salinity, drought stress is aggravated due to limited water uptake by plants from the soil resulting from greater osmotic pressure to root cell (osmotic pressure of soil solution > osmotic pressure in plant's cell sap). Oxidative damage due to soil salinity include detrimental impact on protein, nucleic acid and certain enzymes of plant as there is synthesis of ROS [38]. Under soil salinity, even if

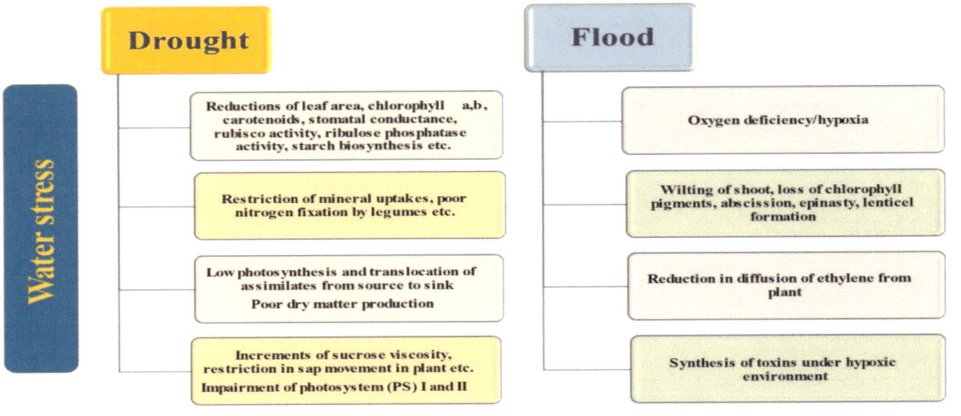

Figure 3.
Negative impacts of water stress (drought and flood) on plants.

uptake of water takes place, there is also intrusion of various salts (Na^+, Cl^-, etc.) inside the plant along with water, which exert negative impact on plant's cell through by impairing activities of various essential enzymes. Plants show burnt like visual symptoms on leaves under excessive salt uptake. Salt stress not only increases the Na^+, Cl^-, etc., but also causes deficiency of various essential elements like calcium (Ca^{2+}), potassium (K^+), magnesium (Mg^{2+}), nitrate (NO_3^-), etc. in rhizospheric zone of soil. Calcium (Ca^{2+}), potassium (K^+), magnesium (Mg^{2+}), nitrate (NO_3^-) are known to influence photosynthesis and therefore, their limited uptake by plants under soil salinity results in less photosynthesis and translocation of assimilates from source to sink. Some major impacts of soil salinity include less leaf expansion, stunted growth, less dry weight of plant, sterility of florets, loss of pollen viability, high epidermal thickness, mesophyll thickness, palisade cell length and diameter, spongy cell diameter, reduction of intercellular space in leaves of plant [7]. Partial or complete of stomata under high salt situation causes less transpiration and cell division resulting in reduction in plant's growth, defoliation and senescence of aerial parts and eventually, plant dies [39]. Under salinity stress, Na^+/K^+ ratio of the cell is excessively increased, resulting in reduction in cell turgidity, enzyme activity and membrane potential of plant. Further, due to abundance of Na^+ in cell, various essential enzymatic activities get downregulated resulting in impairments of cell expansion as well as division, membrane stability and cytosolic metabolism.

3.4 Temperature stress

High temperature or heat stress increases evapotranspiration loss of water resulting in drought like situation. This is further triggered under increase of soil temperature coupled with drought. Due to high temperature, respiration exceeds photosynthesis resulting in depletion of food reserve or loss of carbon (respiration rate doubles with each 10°C rise in tissue temperature). It is also observed that sudden temperature rise causes relatively more harm than gradual increase in temperature due to higher reductions of biochemical, physiological and molecular activities of the plant by sudden temperature rise. Among the categories of plant, C_3 plants suffer comparatively more than C_4 plants due to fluctuations in energy supply and carbon metabolisms under high temperature (**Figure 4**) [40].

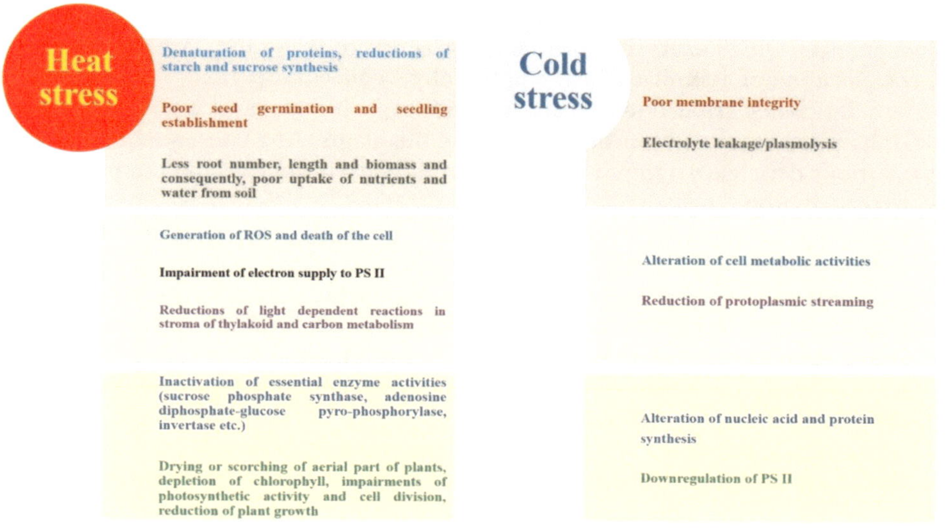

Figure 4.
Negative impacts of temperature stress (heat and cold) on plants.

On the other hand, cold stress or low temperature causes chilling and freezing injuries to the plant. Chilling injury results in disfunctioning of physiological properties, while freezing injury results in cell dehydration. Some major impacts of cold stress on plant include Wilting, bleaching through pigment photo-oxidation, leaf necrosis, browning, cell death, etc.

3.5 Heavy metal toxicity

When there is abundance of heavy metals in soil, plant's physiological, morphological, biochemical, molecular activities are highly affected. After being taken up by the plant's roots, these metals (Pb, Cu, Hg, etc.) move inside the plant through xylem due to transpiration pool and negatively impact nutrient distribution, photosynthesis, enzyme activities, Cu/Zn-SOD, ethylene receptors, etc. resulting in reduction of molecular oxygen content and increment of ROS [41]. Synthesis of ROS thereafter, damages the plant at cellular level.

3.6 Other abiotic stresses

Chemical toxicity/persistence in environment is a great concern today, which results from excessive and unscientific application of chemicals. Environmental hazard or pollution under chemical toxicity leads to poor ecosystem health and diversity. These chemicals not only include pesticides but also cover inorganic fertilizer. Unnecessary use of pesticides and fertilizers are creating climate change resulting in reductions of crop growth, yield and quality. Specifically, complete dependence on chemicals for crop production results in damage of soil health and eventually, soil productivity declines. Contamination of underground fresh water as well as surface water, air pollution, land pollution, etc. is commonly associated with chemical toxicity. Plants known to be grown in an area earlier, are facing trouble in adaptation to changing climate in the same area. Changing climate is linked with various biotic and abiotic stresses which exert detrimental impacts on crop's germination, photosynthesis, translocation of assimilates, etc. and thereby, reduces crop yield. On a contrary, nutrient deficiency arises due to scarcity of nutrients in soil, resulting in their less uptake by plant roots and translocation inside the plants. As a consequence, plants show various deficiency symptoms and its growth diminishes, leading to poor yield and quality of crop.

When wind blows at high velocity over the crop field, plants specially the tall growing or weaker one lodges down, resulting in poor growth, shedding of flower, pollen, grains and thereby, reduction in yield. High wind velocity also causes soil erosion and washes the essential nutrients away from plants. Further, there is an increase in evapotranspiration loss of water under high wind velocity, which demands for frequent water application leading to high cost of cultivation and failure of supply of water leads to poor growth and yield of crop.

Light is one of the prime requisites for photosynthesis and therefore excessive light can disrupts photosynthetic apparatus (photoinactivation and photodamage), while scarcity of light reduces photosynthesis and dry matter production. Plant's growth reduces when light is less or shading by taller plants/trees or other structures occurs. Due to hot sunlight intensity, heat stress or drought stress occurs which alone or together, affects the crop growth. UV ray impairs DNA and causes leaf bleaching, oxidative stress through synthesizing ROS. Under excessive light, breakdown of D1 protein of PS II and decrement of PS I polypeptides like PsaA, PsaB, and PsaC occurs in plants [42].

4. Organic farming and its components

Sustainable agriculture greatly relies on non-chemical, eco-friendly organic farming approaches. Organic farming is defined as holistic production management system which promotes and improves agro-ecosystem health covering bio-diversity, biological cycle and soil biological properties. It completely or largely excludes the use of synthetic off-farm inputs like fertilizers, pesticides, growth regulators, livestock feed additives, etc. and mostly relies on on-farm agronomic, biological and mechanical inputs such as crop rotations, crop residues, organic manures, biofertilizers, green manuring, organic wastes, mineral grade rock additives, biological means of nutrient mobilization and plant protection (botanical pesticides), etc. leading to improvement of soil health, crop growth and yield as well as safety of environment. The major components of organic farming are briefly highlighted below.

4.1 Organic manures and biochar

Organic manures are the sources of nutrients produced by decomposition of organic waste materials (crop residues, plant-based wastes from house/farm/market, etc., animal-based wastes like urine, dung, litter, excreta, etc.) through microbial actions. These are known as bulky organic manures (FYM, vermicompost, poultry manures, common compost, night soil, sewage and sludge, kitchen compost, etc.) as their requirements are high. Besides, there are concentrated organic manures like oilcakes, bone meal, blood meal, horn and hoof meals, fish meal, meat meal, etc. in which more nutrients are present and they, therefore, supply different nutrients relatively in large quantities from unit quantity applied than bulky ones. Apart from solid organic manures, there are various organic liquid manures/ITK formulations such as *Jiwamrit*, *Beejamrit*, *Amrit pani*, *Kunapajala*, *Panchagavya*, *Sanjeevani*, etc. are also used to improve soil health and thereby, crop growth.

Biochar is an excellent soil ameliorant produced under high temperature through controlled pyrolysis of organic substances. Quality of biochar depends on feedstock, temperature and pyrolysis conditions and time. Application of biochar improves plant growth and yield by reviving soil health.

4.2 Crop rotation and other agronomic practices

Crop rotation involves diversification of crops, that is, growing of different crops in succession on same field to avoid pest, disease and weed infestation, improve soil fertility, recycle nutrient reserves, utilize different resources properly, enhance crop productivity, profitability, etc. Besides, there are various other agronomic practices such as variety selection, land preparation, mulching, crop residue retention on soil surface, manure application, time and method sowing, seed rate, spacing and depth, physical, cultural or biological methods of weed, pest and disease control, timely and adequate irrigation, timely harvest and post-harvest operations, followed in organic farming to enhance crop productivity, quality and profitability in production.

4.3 Crop residue

Crop residue is the remaining left after harvesting and separating the economic part from the entire plants. These residues are often burnt leading to environmental pollution. There can be multiple uses of these crop residues like mulching materials, livestock feed, raw materials for manure preparation, substrates for mushroom cultivation, roof thatching, etc. Crop residues can conserve soil moisture, reduce weed infestation and promote crop protection.

4.4 Bio-fertilizers

These are the substances containing living organisms, that is, micro-organisms which are helpful for crop growth and productivity by improving soil health and fertility as well as uptake of nutrients and water by the plants. Seed inoculation or soil application of biofertilizer containing various bacteria (rhizobium, azotobacter, azospirillum, etc.), fungi (VAM, AMF, *Penicillium* sp., *Aspergillus awamori*, etc.), azolla, blue green algae, etc. can help the crop growth either by drawing nutrients and water near rhizospheric zone from distant area or by fixing atmospheric nitrogen in soil leading to enhancement of soil fertility.

4.5 Bio-pesticides and other protection measures

Bio-pesticides such as nicotine, pyrethrum, rotenone, subabilla, ryanin, margosa, neem, etc. are natural plant-based products containing secondary metabolites like alkaloids, terpenoids, phenolics and minor secondary chemicals. Besides, resistant variety selection, myco-pesticides, release of natural enemies or growing trap crop or plants which act as host for biocontrol agents can protect crops from disease and pest damages. Further, various agronomic approaches like mulching, soil solarization, stale seed bed technique, timely and line sowing, crop rotation, intercropping, smother crops, use of botanical extracts, etc. can suppress weed problem.

5. Different organic farming approaches to mitigate abiotic stresses

Over the years, organic farming has served as an eco-friendly approach to improve agricultural productivity in a sustainable manner. Further, it acts as buffer against various biotic and abiotic stresses which are often less highlighted. In the following section, different organic farming practices having the potential to mitigate various abiotic stresses are mentioned.

Mulching: In organic farming, mulching with straw, compost and other crop residues plays a key role in mitigating drought stress along with associated salt accumulation on soil surface. Further, it reduces the chances of loss of surface soil nutrients by restricting the soil erosion arising due to direct impact of rainfall or high runoff velocity. Mulch materials also act as insulators which keep the soil cool during warm weather and hot during winter months and thereby, solving the issues of heat and cold injuries to a high extent. Decrease of soil temperature by 1–2°C has been reported by Král et al. [43]. Apart from their role in soil and moisture conservation and checking different abiotic stresses, mulch materials like crop residues, compost, etc. can improve soil microbial activities and add essential nutrients through decomposition over the time [43]. In reality, abiotic stresses are most often associated with biotic stresses. Mulching, beside alleviation of various abiotic stresses, also suppresses various weed infestations in crop field. All these benefits are directly reflected to high crop growth, yield and quality under climate change scenario.

Sea weed extracts: Sea weed extracts are now emerging as one of potential sources of nutrients in organic farming for crop production. They contain nutrients, plant growth promoting substances, enzymes as well as antioxidants which help the crop cope up with salinity, heat and drought stresses. Besides, the use of sea weed extracts for cold tolerance as well as associated nutrient deficiency is now emerging. Algal extracts trigger a number of pathways to enhance stress tolerance through scavenging ROS. These extracts improve soil properties to conserve water well and thus, allow crop to survive under drought. Earlier, the use of these extracts was successful on Kentucky bluegrass (*Poa pratensis* L. cv. Plush) to mitigate salinity stress also [44].

Organic manures and biochar: Organic manures as well as green manure plants contain various nutrients, growth regulators, micro-organisms, etc. which not only improve soil fertility by solving nutrient scarcity stress but also improve overall soil health to a high extent. Increasing soil porosity, aggregate stability, reduction of compactness, etc. increase water holding capacity of the soil and thereby, address the issues of drought stress. Further, moderation of soil temperature, greater aeration in root zone, beneficial micro-organisms' activities in soil improves crop growth and yield. It has been widely noted that organic matter through decomposition releases humic and fulvic acids which alleviate abiotic stresses.

Apart from the organic manures, biochar application can alleviate various abiotic stresses specially drought stress. Shashi et al. [45] observed positive result on maize from rice husk @ 20 t/ha biochar under drought condition by enhancing bacterial and fungal communities in soil. Biochar specially from poultry manure shows excellent properties to mitigate salinity stress by reducing Na and increasing CEC and SOC contents in soil. Further, biochar can protect plants from high and low temperatures as well as alleviates metal toxicity by immobilizing heavy metals, followed by reducing their mobility. Positive impact of biochar in mitigating different abiotic stresses in rice is summarized in **Table 2**.

Biofertilizers and bio-stimulants: Biofertilizer is known to improve soil fertility and overall soil health through accelerating beneficial micro-organisms' activities. Besides,

Abiotic stresses	Biochar type	References
Acidity	Sewage sludge	[46]
Salinity	Bamboo	[47]
Salinity	Rice husk	[48]
Nutrient deficiency	Rice straw	[49]
Saline-sodic stress	Wheat straw	[50]
Saline-sodic stress	Groundnut shell	[51]
Cold stress	Bamboo	[52]
Nutrient deficiency	Rice residue	[53]
Cold stress	Bamboo	[54]
Nutrient deficiency	Rice husk	[55]
Heat stress	Rice husk	[56]
Heat stress	Rice husk	[57]
Cd stress	Rice and maize residues	[58]
Cd and Pb stresses	Wheat straw	[59]

Table 2.
Positive impact of biochar in mitigating different abiotic stresses in rice.

biofertilizer is one of the key components of organic farming to alleviate various abiotic stresses. Various types of biofertilizer helps the crop to tolerate or overcome stresses resulting in good growth and productivity under stress situation. It has been also found that seed bio priming with micro-organisms alleviates various abiotic stresses through improving germination and early plant stand establishment [60]. Bio priming increases the osmolyte concentrations leading to high cell wall elasticity and turgid weight to dry weight ratio. Further, endophytic synthesis of alkaloids protects macromolecules through ROS scavenging activities. Plant growth promoting rhizobacteria (PGPR) improves drought responsive genes' expression through high ROS scavenging activities. It also synthesizes phytohormones like IAA, GA3, etc. resulting in high plant growth under stress. PGPR also synthesizes exopolysaccharides resulting in good soil structure and uptakes of nutrients and water. Various endophytic micro-organisms also confer abiotic stress tolerance in plants through activating host stress response as well as through synthesizing biochemicals against stresses. A specific category of microorganisms known as arbuscular mycorrhizal fungi (AMF) is well known to mitigate negative impacts of various abiotic stresses on crop by improving soil health and plant's defense mechanism. It makes symbiotic relationship with roots of around 90% of the plant types. Use of AMF as biofertilizer/bio-inoculant is an emerging strategy specially under climate change scenario. The fungal network extends as secondary root system and helps the crop to draw nutrients and water from distant areas. Further, it plays a key role in regulating anti-oxidant activities (CAT, POX, SOD, GST, etc.) of plants under specific or combined stress situation resulting in scavenging of ROS and improvement crop growth, yield and quality. Various stress alleviating properties of micro-organisms in the form of biofertilizers/priming are shown in **Table 3**.

Bio-stimulants are organic or inorganic substances rich in bioactive compounds and/or micro-organisms, which improve crop growth through developing root for

Microorganisms	Crop	Abiotic stress	Impacts	References
Sinorhizobium meliloti	Rice	Temperature stress	High endogenous hormone and photosynthesis	[61]
Rhizobium trifolii	Berseem	Salinity stress	High dry matter and nodulation	[62, 63]
Pseudomonas putida	Cotton	Salinity and alkalinity stress	High seed germination, plant height, fresh and dry weights through increased uptake of K^+, Mg^{2+} and Ca^{2+} and decreased uptake of Na^+	[64]
Pseudomonas alcaligenes, Bacillus polymyxa, Mycobacterium phlei	Maize	High temperature and salinity stresses	Calcisol produced by bacteria	[65]
Achromobacter piechaudii	Tomato, pepper	Salinity and water stresses	High biomass production	[66]
Pseudomonas mendocina	Lettuce	Salinity stress	High shoot production	[67]
Azospirillium spp., *Phosphobacteria* spp., *Glucanacetobacter* spp.	Rice, mangroves	Heavy metal (Fe) toxicity	Improvement in crop growth	[68, 69]
P. putida, Pseudomonas fluorescens	Canola, barley	Cd toxicity	Enhancement in IAA, siderophore and 1-aminocyclopropane-1-carboxylate deaminase	[70]
Pseudomonas sp.	Basil	Water stress	High antioxidant and photosynthetic pigments	[71]
Arbuscular mycorrhiza	Rice	Drought stress	High antioxidant and photosynthetic efficiency	[72]
Inoculation with AMF and PGPR	Date palm	Drought stress	High proline content and relative water content; low SOD, CAT, GST and POX activities in leaf	[73]
Glomus intraradices, P. mendocina	Different plants	Drought stress	Low SOD activity	[74]
Kluyvera ascorbata	Tomato	Ni, Pb, Zn toxicity	Reduction in accumulation of heavy metals in plants	[75]
Methylobacterium oryzae, Burkholderia sp.	Tomato	Ni, Cd toxicity	Less uptake and translocation of heavy metals	[76]
Pseudomonas brassicacearum, Pseudomonas marginalis, Rhodococcus sp.	Pea	Nutrient deficiency	Stimulation of root and high nutrient uptake	[77]
Rhizobium sp.	Pea	Zn and Ni toxicity	Enhancement of plant growth parameters	[78]
Arbuscular mycorrhizal fungi	Olive	Drought stress	High turgor potential and mineral nutrient uptake	[79]
Arbuscular mycorrhizal fungi	Soybean	Drought stress	High leaf area index, photosynthesis, growth and yield	[80]

Microorganisms	Crop	Abiotic stress	Impacts	References
Rhizophagus irregularis	Pangola grass	Drought stress	High stomatal conductivity, low lipid peroxidation	[81]
Glomus mosseae	Wheat	Drought stress	High chlorophyll, osmotic potential, antioxidant activities	[82]
Glomus etunicatus	Onion	Drought stress	High fresh and dry matter, phosphorus content	[83]
G. intraradices	Tomato	Salinity stress	High ion uptake, chlorophyll, growth and dry matter	[84]
R. irregularis	Tomato	Salinity stress	High root, shoot, leaf number, growth hormone synthesis	[85]
Claroideoglomus etunicatum	*Aeluropus littoralis*	Salinity stress	Stomatal conductance, root and shoot dry matter, sugar content	[86]
Glomus fasciculate	*Acacia nilotica*	Salinity stress	High root and shoot dry matter, Zn, Cu, P uptakes	[87]
G. mosseae	Cucumber	Salinity stress	High biomass, synthesis of antioxidant enzymes and photosynthesis pigments	[88]
R. irregularis, Glomus versiforme	Barley	Temperature stress	High survival rate	[89]
Funneliformis sp.	Maize	Temperature stress	Maintenance of PS II heterogeneity	[90]
R. irregularis	Cucumber	Temperature stress	High photosynthetic rate	[91]
Funneliformis mosseae	*Elymus nutans*	Temperature stress	High plant growth, chlorophyll and antioxidants, low oxidative damage	[92]
Glomus isolates	Maize	Heavy metal stress	High Mg, P and K contents in plants, dry matter	[93]
F. mosseae	*Trigonella foenum-graceum*	Heavy metal stress	High crop growth and yield	[94]
G. versiforme	*Lonicera japonica*	Heavy metal stress	Low Cd content in root and shoots	[95]
R. irregularis	*Populus alba*	Heavy metal stress	Low Zn and Cu toxicity	[96]
G. mosseae	*Trifolium pratense*	Heavy metal stress	Low root and shoot concentrations, Zn uptake	[97]
Acaulospora trappei, Glomus leptotichum	*Panicum hemitomon*	Flood stress	High P content in plant	[98]
G. intraradices	*Pterocarpus officinalis*	Flood stress	High growth and P content in leaves	[99]
Glomus geosporum	*Aster tripolium*	Flood stress	High sugar and proline content	[100]

Table 3.
Micro-organisms in the form of biofertilizers/priming against abiotic stresses.

high absorption and assimilation efficiency of nutrients, regulating proper water balance in plants as well as tolerating various abiotic stresses by synthesizing proline, simple sugars, alcohols, abscisic acid, osmotic compounds and antioxidants (to scavenge ROS) [101]. Role of bio-stimulants in plants is shown in **Figure 5**. It increases the contents of carotenoids, phenolic compounds and other secondary metabolites in plants as defense against stresses. It is applied as soil drench (directly/through irrigation) or foliar spray or treatment of seeds. Mitigation of various abiotic stresses by bio-stimulants is listed in **Table 4**.

Crop rotation and various agronomic interventions: Crop rotation is one of the key principles of conservation agriculture. It is always suggested to add leguminous crop in rotation to revive soil fertility after cultivation of a soil exhaustive crop through fixing atmospheric nitrogen. Further, biomass incorporation in soil results in addition of SOC content and thereby, causes improvement of soil porosity, water holding capacity, soil fertility, etc. leading to protection of plants against drought, salinity, high temperature stress as well as nutrient deficiency. Growing a shallow rooted crop after deep rooted crop helps in utilization of nutrients and water from various depths of soil profile so that plant can't experience nutrient and water scarcity.

Various other agronomic practices also can protect the crop from being affected by abiotic stresses under climate change scenario (**Figure 6**, **Table 5**). For instance, proper selection of resistant/tolerant crop and varieties under a prevalent abiotic stress is one useful strategy. To achieve this, breeding activities should include identification of responsive genes. Grafting is another one, which is widely used in horticulture to counter various abiotic stresses specially, salinity, nutrient or water deficiency, heavy metal toxicity, etc. Here, scion susceptible to stress is grafted to stress tolerant root stock. Exogenous application of plant components such as amino acid, sugars, etc. and phytohormones such as ABA, GA3, jasmonic acid, salicylic acid, brassinosteriods, etc. protects crop from abiotic stresses. Application of citric acid and vitamin C exhibit antioxidant properties which inactivates heavy metals such as Cu, Pb, Al, etc. as well as helps crop to overcome salinity and drought stresses through ROS scavenging activities. Soil and foliar applications of humic substances, beneficial fungi, bacteria, chitosan, sea weed extracts, etc. can combat abiotic stresses. Tillage also plays key role in conserving moisture and nutrients as well as breaking hard pan and high percolation of water and thereby, mitigates drought, flood and salinity. Keeping the land fallow for a season or year can rejuvenate the soil fertility and

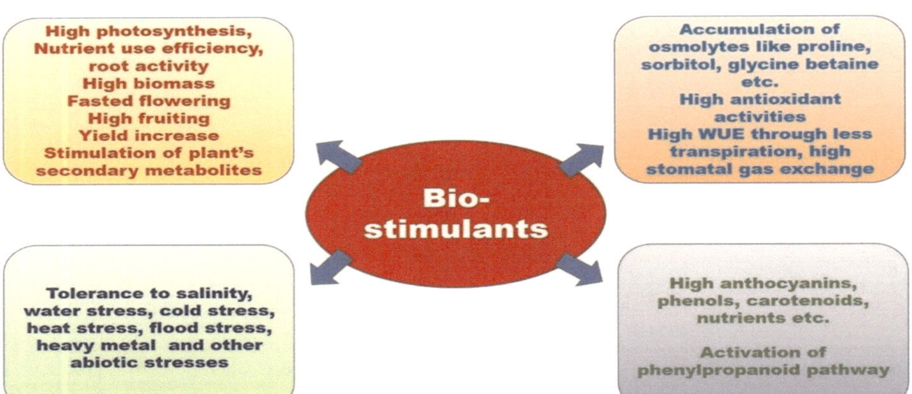

Figure 5.
Role of bio-stimulants in plants.

Abiotic stress	Crop	Bio-stimulants	Impacts	References
Cold stress	Coriander	Asahi SL @0.1%	High chlorophyll *a* and carotenoids	[102]
	Tomato	*Flavobacterium glaciei, Pseudomonas frederiksbergensis, Pseudomonas vancouverensis*	High shoot and root length and biomass, low electrolyte leakage, lipid peroxidation, proline accumulation, SOD, CAT, APX , POD and GR activities	[102]
	Strawberry	Pepton 85/16 @ 2 L/ha or 4 L/ha	New root initiation, more flowering and fruiting	[103]
	Lettuce	Pepton 85/16 @ 04, 0.8, 1.6 g/L	High fresh and dry weights, relative growth rate	[104]
	Lettuce	Terra-Sorb Foliar	High root fresh weight, green cover %	[105]
	Chilli	5-Aminolevulinic acid	High chlorophyll, relative water content, shoot and root biomass, SOD activity, low membrane permeability	[106]
	Pepper	*Serratia nematodiphila*	High plant growth and regulation of endogenous GA4, abscisic acid, jasmonic acid and salicylic acid	[107]
Drought stress	Tomato	Megafol @ 2 ml/L	Increased leaf area	[108]
	Spinach	*Ascophyllum nodosum* @0.50%	Increased leaf area, fresh and dry weights	[109]
	Pea	*P. putida, P. fluorescens*	High root and shoot lengths, pods/ plant, chlorophyll content and grain yield	[109]
	Tomato, chilli	*A. piechaudii*	Low ethylene synthesis, high fresh and dry weights of seedling	[110]
	Tomato	*A. nodosum* @0.33%	High relative water content, plant growth, foliar density, proline and sugar contents, chlorophyll content, low lipid peroxidation	[111]
	Mustard	*A. nodosum* + amino acids	Increase in chlorophyll activity	[112]
	Tomato	VIVA	High shoot and root biomass	[113]
	Basil	*Pseudomonades, Bacillus lentus, Azospirillum brasilens*	High CAT, GPX and chlorophyll activities	[114]
	Pumpkin	Moringa leaf extract	High growth, harvest index, water use efficiency, low electrolyte leakage	[115]
	Soybean	*Sphingomonas* sp.	High growth, chlorophyll content, amino acid, sugar, low ABA and JA	[116]
	Wheat	*Bacillus subtilis*	Enhancement in IAA, decrease in ABA	[117]
	Lucerne	*Enterobacter ludwigii* and *Bacillus megaterium*	Low electrolyte leakage, ABA level, high growth, chlorophyll content, relative water content, nutrient concentrations	[118]

Abiotic stress	Crop	Bio-stimulants	Impacts	References
	Foxtail millet	*P. fluorescens*	ACC deaminase production, high seedling growth	[119]
	Potato	*B. subtilis*	High tuber weight, soluble sugar and CAT, POD, SOD activities	[120]
	Tobacco	Arbuscular mychorrhizal fungi and PGPR	High growth, chlorophyll content, phenol and flavonoid levels	[121]
	Finger millet	*P. fluorescens, P. palleroniana*	High plant growth, nutrient contents, leaf pigment and proline contents, SOD, CAT, GPX activities, low lipid peroxidation	[122]
	Maize	(*Bacillus pumilus, Bacillus cereus*, *Pseudomonas* sp., and *Proteus* sp. and protein hydrolysates	High ABA, IAA, GA, relative water content, protein content, photosynthetic pigments	[123]
	Rice	Commercial seaweed extract (*A. nodosum*)	High plant biomass, yield, leaf area index, chlorophyll content	[124]
	Tomato	Protein hydrolysate	High plant growth, pollen viability, leaf water potential, lycopene content	[125]
Flood stress	Sesame	*Pseudomonas veronii*	High fresh and dry biomass, root and shoot length, chlorophyll content	[126]
	Wheat	*Trichoderma asperellum*	Low ethylene synthesis, high seedling growth	[127]
Heat stress	Common bean	Brassinosteroids @ 25, 50 and 100 ppm	High plant growth, leaves, branches and shoots/plant, fresh and dry weights, nutrient contents	[128]
	Green gram	Glutathione @ 0.5 Mm	High chlorophyll, proline contents, low ROS	[129]
	Chick pea	Proline @ 5, 10, 15 Mm	High germination, shoot and root lengths, proline content, chlorophyll synthesis, low electrolyte leakage, lipid peroxidation	[130]
	Chick pea	Abscisic acid @ 2.5 Mm	High shoot length, chlorophyll content	[131]
	Tomato	*B. cereus*	High growth, APX, SOD, GSH activities, ion uptake (Fe, P, K), chlorophyll content.	[132]
	Chinese cabbage	*Bacillus tequilensis*	High shoot, leaf developments, JA and salicylic acid production, low ABA	[133]
	Rice	Brassinosteroids, amino acids, nitophenolatres, or botanical extracts	High photosynthesis, stomatal conductance, low lipid peroxidation and proline content	[134]
	Tomato	Commercial seaweed extracts	Improved root system, chlorophyll content, high growth	[135]
Fe deficiency	Strawberry	Actiwave (*A. nodosum*) @ 10 ml/20 ml water	High vegetative growth, chlorophyll content, stomatal density,	[136]

Abiotic stress	Crop	Bio-stimulants	Impacts	References
			photosynthesis, fruit production, berry weight	
NPK deficiency	Tomato	Amino acids @ 0.1, 0.2 ml/L water	High plant growth, root and leaf	[137]
	Okra	Kelpak (*Ecklonia maxima*) @ 0.40%	More number of leaves, roots, stem thickness, shoot weight, root weight, leaf area	[138]
	Garlic	Bio-Cozyme @ 2 kg/ha	High bulb yield, plant height, nutrient in leaves	[139]
Salinity	Lettuce	*A. brasilense*	High germination, seedling growth, chlorophyll, dry mass	[140]
	Chilli	*A. brasilense/Pantoea dispersa*	High plant dry weight, carbon di oxide assimilation, nitrate concentration	[141]
	Pea	*Rhizobium leguminosarum*	High plant growth	[142]
	Pumpkin	*Bacillus pumilis, Trichoderma harzannum, Paenibacillus azotoformans*	High fresh weight, potassium uptake, low sodium uptake	[143]
	Common bean	Humic acid@ 0.05%, 0.1%	High nitrogen and phosphorus, plant root and shoot growth, low electrical conductivity, electrolyte leakage	[144]
	Strawberry	Acadian (*A. nodosum*)	High growth and yield	[145]
	Lettuce	Super Fifty (*A. nodosum*) @ 0.4, 1, 2.5, 10 ml/L	High root, stem and total plant biomass	[146]
	Lettuce	Protein hydrolysates @ 2.5 ml/L	High plant shoot and root growths, fresh yield, low oxidative stress	[147]
	Tomato	*A. piechaudii*	High fresh and dry weights, uptakes of phosphorus and potassium, water use efficiency, low ethylene production	[148]
	Cucumber	*A. nodosum* @ 1, 2 g/kg	High fruit yield	[149]
	Common bean	Licorice root extract @ 0.50%	Plant growth, yield, relative water content, total soluble sugars, low electrolyte leakage	[150]
	Common bean	Propolis and maize grain extract @ 1%, 2%	High germination, seedling growth, proline, total soluble sugars, low electrolyte leakage, ABA, lipid peroxidation	[151]
	Common bean	Moringa oleifera	High shoot and root growths, total soluble sugars, proline, SOD, APX, GR activities	[152]
	Chick pea	*Sargassum muticum* and *Jania rubens* @ 1%	High chlorophyll, carotenoids, plant growth, soluble sugars, CAT, SOD, POD, APX activities, low MDA	[153]
	Tomato	*Dunaliella salina* exopolysaccharides @ 0.1 g/L	High protein, chlorophyll, low proline	[154]

Abiotic stress	Crop	Bio-stimulants	Impacts	References
	Onion	Bee-honey based bio-stimulant @ 25–50 g/L	High water use efficiency, bulb yield, antioxidants, photosynthetic pigments	[155]
	Chilli	Humic acid @ 50, 100, 150 mg/kg	High fresh and dry weights, nutrient uptakes, low membrane damage	[156]
	Pea	*Acinetobacter bereziniae, E. ludwigii, Alcaligenes faecalis*	Low electrolyte leakage, high proline, chlorophyll, total soluble sugar, plant growth	[157]
	Tomato	*Pseudomonas oryzihabitans*	Increase plant growth and photosynthetic characters	[158]
	Ground nut	*Stenotrophomonas maltophilia*	High growth, auxin and total amino acids, low proline, electrolyte leakage, lipid peroxidation	[159]
	Common bean	*Aneurinibacillus aneurinilyticus*	High root and shoot length and weight, chlorophyll content	[160]
	Soybean	*Arthrobacter woluwensis, Microbacterium oxydans, Arthobacter aurescens, B. megaterium, B. aryabhattai*	High antioxidant (SOD, GSH) activities, chlorophyll content	[161]
	Wheat	*Trichoderma reesei*	High plant biomass, chlorophyll, carotenoids, uptake of nutrients, low Na, ABA contents	[162]
	Wheat	*Trichoderma longibrachiatum*	High root and shoot length and weight, relative water content, chlorophyll content, antioxidant (SOD, POD, CAT) activities	[163]
	Soybean	*Porostereum spadiceum*	High seedling growth	[164]
	Cucumber	*C. etunicatum, Rhizophagus intraradices*, and *F. mosseae*	High biomass, SOD, CAT, APX and GR activities, JA, SA contents, low lipid peroxidation, electrolyte leakage	[88]
	Maize	Humic acid	High photosynthesis rate, plasma membrane proton pumps activity	[165]
	Rice	Panchagavya	High plant growth, chlorophyll, carotenoid, anthocyanin contents, low CAT, SOD, POX activities	[166]
	Tomato	Seaweed extract (*Ulva lactuca*)	High plant growth, soluble sugar, total protein, chlorophyll content, carotenoids, low hydrogen peroxide, APX activity	[167]
	Wheat	*Exiguobacterium aurantiacum*	High plant growth, nutrient content, proline contents, CAT and POD activities	[168]
Heavy metal stress	Rice	*E. ludwigii* and *Exiguobacterium indicum*	High growth, chlorophyll content, SA, low ABA, Ni and Cd	[169]
	Cucumber	*Pseudomonas psychrotolerans*	High growth, chlorophyll content, IAA and GA	[170]

Abiotic stress	Crop	Bio-stimulants	Impacts	References
	Soybean	*B. cereus*	High shoot and root growth, SA, chlorophyll content, low ABA, POD activities, low Cd accumulation	[171]
	Sunflower	Arbuscular mycorrhizal fungi	High growth, antioxidant activities, fatty acid contents	[172]

Table 4.
Mitigation of abiotic stresses through various bio-stimulants.

Figure 6.
Abiotic stress mitigation through various organic farming practices.

moisture content for next crop. Timely and properly sowing, adequate seed rate, spacing and depth, seed treatment also allows the crop to grow and utilize resources properly resulting in surviving and withstanding of climate change scenario. For instance, wheat, if sown on time, can escape terminal heat stress. Further, adequate and timely water, nutrient and interculture (weeding) managements accelerate crop growth by conservating water, nutrients, light, etc. which otherwise could be utilized by weeds and thereby, mitigate drought, salinity, nutrient deficiency, etc. Tall variety is susceptible to lodge by high wind velocity, while dwarf, robust variety can withstand the wind stress. Shelterbelt also protects the crop from high wind. Sometimes, crop suffers from hot sunlight and requires shading from tall growing crop and thus, intercropping or agroforestry is beneficial. On a contrary, shading of tall weeds on crop affects crop growth and therefore, timely weed management is needed. Under saline condition, frequent flooding with irrigation water or irrigation to root by drip method, scraping of surface salts, application of plant growth promoting bacteria, etc. are the key mitigation practices. PGPB alleviates salinity through hydraulic conduct, osmotic accumulation, toxic sodium removal, higher osmotic activity. Further, use of organic product such as brewer's spent grain as soil amendment not only improves soil

Abiotic stresses	Agronomic management practices in organic farming	Abiotic stresses	Agronomic management practices in organic farming
Drought	• Use of resistant/tolerant variety • Mulching or cover cropping or inter/mixed cropping to reduce evaporation loss of water (moisture conservation) • Sowing in ridge and furrow bed • Alternate/skip furrow irrigation or partial root drying • Skip row planting • Use of sprinkler/drip or any other micro irrigation/water saving options • Less application of manures • Use of bio-fertilizers and seed priming • Nipping or pinching apical portion to arrest shoot growth and consequently, transpiration rate	Flood	• Use of resistant variety • Drainage of excess water • Growing of water loving crops • Double transplanting in rice • Growing tall plants to avoid complete submergence
		Salinity	• Use of resistant/tolerant variety • Incorporation of green manure crop in soil before sowing • Exogenous applications of non-synthetic ABA and/or jasmonic acid • Seed treatment with non-synthetic polyamines viz., putrescine, spermidine, spermine, etc.
High temperature	• Use of resistant/tolerant variety • Shading on the plant canopy • Use of mulch or residue retention to avoid heat stress at early growth stages • Application of non-synthetic salicylic acid or glycine betaine or ethylene or gibberellic acid • Irrigation on the canopy to restrict sun scorching • Drip irrigation to reduce soil temperature at root zone depth • Timely sowing of winter crops to avoid heat stress during anthesis and seed formation phases	Low temperature	• Use of resistant/tolerant variety • Seed treatment with non-synthetic gibberellic acid or proline • Use of non-synthetic cryoprotectants, ABA, paclobutrazol, uniconzole, etc. • Timely sowing of monsoon crops to avoid terminal cold stress
		Heavy metal toxicity	• Construction of wetlands • Reduction of chemical based intensive farming approach • Substitution of chemicals with biofertilizer, compost and bio-pesticides
		Wind velocity	• Use of windbreaks/shelterbelts • Use of dwarf crop varieties
Low light	• Use of sun loving or tall varieties	Excess chemicals and nutrients	• Promotion of organic farming practices • Growth of nutrient exhaustive crops
High light	• Use of shade loving or dwarf varieties	Nutrient scarcity	• Application of nutrients through manures to correct the deficiency

Table 5.
Agronomic management practices in organic farming to mitigate specific abiotic stress.

fertility but also alleviates heavy metal, nutrient deficiency, salinity, drought stresses, etc. Intercropping/Mixed cropping also conserves soil and water, suppresses weeds, reduces salt accumulation on surface through evaporation and thereby, alleviates various stresses. Sometimes, allelopathic potential of many crops on weeds are utilized to suppress weeds resulting in conservation of resources and good crop growth. Under the scarcity of water, precise and wise use of water, clipping of leaves (to reduce transpiration water loss), organic anti-transpirants (like wax, panchagavya) application, broadcasting of seeds, closer spacing, more plant population/hill, double transplanting, etc. are useful. Apart from drainage, double transplanting is also beneficial for flood condition where main field is too flooded to transplant seedlings on time. It is well known fact that various biotic stresses like pest, disease and weeds trigger abiotic stresses. Addressing these biotic stresses by botanical extracts, biopesticides, release of natural enemies or living organisms, trap cropping, etc. can help the crop to avoid various abiotic stresses.

6. Conclusion

Abiotic stress is creating detrimental effect on living organisms specially on plants since long. Its negative impact on crop is becoming prominent in recent days in the context of climate change scenario. In most of the cases, an abiotic stress combines with other abiotic or biotic stresses to exert combined impact on crop growth, yield and quality and the extent of impact on crop varies from mild to severe resulting in hampering crop growth accordingly. Although plants adopt some internal defensive mechanisms to counter these stresses, in most of the times, they require external stimuli/practices/inputs to mitigate abiotic stresses. Due to population rise, crop yield loss through abiotic stresses cannot be accepted at this moment or future and therefore, suitable agronomic and breeding interventions are highly needed. Since chemical-based farming is a barrier against sustainable agricultural production as it deteriorates soil health and is hazardous to the environment due to toxic chemical footprint, organic farming is emerging as its potential alternative. Various organic farming inputs such as organic manures, biofertilizers, bio-priming with micro-organisms, bio-stimulants (seaweed extracts, humic acid, micro-organisms etc.), mulches, biochar etc. have the potential to mitigate abiotic stresses under climate change scenario. Further, organic farming practices like crop rotation, inter cropping, tillage, time and method of sowing, nutrient, water and intercultural operations, use of PGPB, organic formulations, grafting, selection of resistant/tolerant varieties and other scientific/wise uses of organic inputs can help the crop to mitigate/escape the detrimental effects of various abiotic stresses to a great extent. Still, there is need on proper research or study on the abiotic stress potential of organic farming further. Available organic farming technologies as well as information/awareness about them are very also scanty at this moment. Therefore, proper multi-locational research experiments, transfusion of modern practices/awareness through strong extension services, policy interventions and advanced breeding approaches are highly required to address harmful abiotic stresses as well as to get high crop growth, yield and quality. Various strategies should be jointly implemented rather than using individually to get the best result from organic farming in making crop to cope up successfully with climate change scenario.

Author details

Saikat Biswas[1], Rupa Das[2]* and Lay Lay Nwe[1]

1 Department of Agronomy, Bidhan Chandra Krishi Viswavidyalaya, Mohanpur, Nadia, West Bengal, India

2 Department of Seed Science and Technology, Bidhan Chandra Krishi Viswavidyalaya, Mohanpur, Nadia, West Bengal, India

*Address all correspondence to: rupadasbiswas18@gmail.com

IntechOpen

© 2023 The Author(s). Licensee IntechOpen. This chapter is distributed under the terms of the Creative Commons Attribution License (http://creativecommons.org/licenses/by/3.0), which permits unrestricted use, distribution, and reproduction in any medium, provided the original work is properly cited.

References

[1] Chaves MM, Maroco JP, Pereira JS. Understanding plant responses to drought- from genes to the whole plant. Functional Plant Biology. 2003;**30**: 239-264

[2] Rao IM, Beebe SE, Polania J, Ricaurte J, Cajiao C, Garc'ıa R, Rivera M. Can tepary bean be a model for improvement of drought resistance in common bean? African Crop Science Journal. 2013;**21**:265-281

[3] Prasad PVV, Pisipati SR, Momcilovic I, Ristic Z. Independent and combined effects of high temperature and drought stress during grain filling on plant yield and chloroplast protein synthesis elongation factor (EFTu) expression in spring wheat. Journal of Agronomy and Crop Science. 2011;**197**: 430-441

[4] Acquadh. Principle of Plant Genetics and Breeding. Oxford: Willey Blackwell; 2007

[5] Kumar S. Abiotic stresses and their effects on plant growth, yield and nutritional quality of agricultural produce. International Journal of Food Science and Agriculture. 2020;**4**(4): 367-378

[6] Fahad S, Bajwa AA, Nazir U, Anjum SA, Farooq A, Zohaib A, et al. Crop production under drought and heat stress: Plant responses and management options. Frontiers in Plant Science. 2017;**29**:8

[7] Das R, Biswas S. Influence of abiotic stresses on seed production and quality. In: Seed Biology Updates. IntechOpen; 2022. pp. 1-24

[8] Bockheim JG, Gennadiyev AN. The role of soil-forming processes in the definition of taxa in soil taxonomy and the world soil reference base. Geoderma. 2000;**95**(1–2):53-72

[9] Abrol IP. Salt-affected soils: An overview. In: Chopra VL, Paroda SL, editors. Approaches for Incorporating Drought and Salinity Resistance in Crop Plants. New Delhi: Oxford and IBH Publishing Company; 1986. pp. 1-23

[10] Levitt J. Responses of plants to environmental stresses. Academic Press. 1980;**1**:496

[11] Wuana RA, Okieimen FE. Heavy metals in contaminated soils: A review of sources, chemistry, risks and best available strategies for remediation. ISRN Ecology. 2011;**2011**:1-20. DOI: 10.5402/2011/402647

[12] Farooq M, Nadeem F, Gogoi N, Ullah A, Alghamdi SS, Nayyar H, et al. Heat stress in grain legumes during reproductive and grain-filling phases. Crop & Pasture Science. 2017;**68**: 985-1005

[13] Bethke PC et al. Tuber water and pressure potentials decrease and sucrose contents increase in response to moderate drought and heat stress. American Journal of Potato Research. 2009;**86**:519

[14] Kang JS, Singh H, Singh G, Kang H, Kalra VP, Kaur J. Abiotic stress and its amelioration in cereals and pulses: A review. International Journal of Current Microbiology and Applied Sciences. 2017;**6**(3):1019-1045

[15] Singh S et al. Effect of water stress at different stages of grain development on the characteristics of starch and protein of different wheat varieties. Food Chemistry. 2008;**108**:130-139

[16] Di Caterina R et al. Influence of salt stress on seed yield and oil quality of two sunflower hybrids. Annual Applied Biology. 2007;**151**:145-154

[17] Liu Y, Li J, Zhu Y, Jones A, Rose RJ, Song Y. Heat stress in legume seed setting: Effects, causes, and future prospects. Frontiers in Plant Science. 2019;**10**:938

[18] Vaz Patto MC, Amarowicz R, Aryee ANA, Boye JI, Chung HJ, Martín-Cabrejas MA, et al. Achievements and challenges in improving the nutritional quality of food legumes. Critical Reviews in Plant Sciences. 2015;**34**:105-143

[19] Dong B, Zheng X, Liu H, Able JA, Yang H, Zhao H, et al. Effects of drought stress on pollen sterility, grain yield, abscisic acid and protective enzymes in two winter wheat cultivars. Frontiers in Plant Science. 2017;**8**(June):1-14

[20] Sekhon HS et al. Water use efficiency under stress environments. In: Climate Change and Management of Cool Season Grain Legume Crops. The Netherlands: Springer; 2010. pp. 207-227

[21] Ni Z, Li H, Zhao Y, Peng H, Hu Z, Xin M, et al. Genetic improvement of heat tolerance in wheat: Recent progress in understanding the underlying molecular mechanisms. Crop Journal. 2018;**6**(1):32-41

[22] Asseng S, Foster I, Turner NC. The impact of temperature variability on wheat yields. Global Change Biology. 2011;**17**(2):997-1012

[23] Kazai P, Noulas C, Khah E, Vlachostergios D. Yield and seed quality parameters of common bean cultivars grown under water and heat stress field conditions. AIMS Agriculture and Food. 2019;**4**:285-302

[24] Wilhelm I. Crop physiology and metabolism. Crop Science. 1999;**39**: 1733-1741

[25] Jagadish S, Craufurd P, Wheeler T. High temperature stress and spikelet fertility in rice (*Oryza sativa* L.). Journal of Experimental Botany. 2007;**58**(7): 1627-1635

[26] Redden RJ, Vara HJL, Prasad PV, Ebert AW, Yadav SS, O'Leary GJ. Temperature, climate change, and global food security. Temperature and Plant Development. 2014;**8**:181-202

[27] Vafa P et al. The effect of drought stress on grain yield, yield components and protein content of durum wheat cultivars in Ilam Province, Iran. International Journal of Agricultural and Biosystems Engineering. 2014;**8**:631-636

[28] Wardlaw IF et al. Contrasting effects of chronic heat stress and heat shock on kernel weight and flour quality in wheat. Functional Plant Biology. 2002;**29**:25-34

[29] Triboï E et al. Environmentally induced changes in protein composition in developing grains of wheat are related to changes in total protein content. Journal of Experimental Botany. 2003; **54**:1731-1742

[30] Flagella Z et al. Changes in seed yield and oil fatty acid composition of high oleic sunflower (*Helianthus annuus* L.) hybrids in relation to the sowing date and the water regime. European Journal of Agronomy. 2002;**17**:221-230

[31] Lin CJ et al. Influence of high temperature during grain filling on the accumulation of storage proteins and grain quality in rice (*Oryza sativa* L.). Food Chemistry. 2010;**58**:10545-10552

[32] Sadeghipour O. The influence of water stress on biomass and harvest

index in three mung bean (*Vigna radiata* L. (Wilczek)) cultivars. Asian Journal of Plant Sciences. 2009;**8**:245-249

[33] Silva JAB et al. Microtuberization of Andean potato species (*Solanum* spp.) as affected by salinity. Scientia Horticulturae. 2001;**89**:91-101

[34] Dornbos DL, Mullen RE. Soybean seed protein and oil contents and fatty acid composition adjustments by drought and temperature. Journal of the American Oil Chemists' Society. 1992;**69**: 228-231

[35] Hoegy P et al. Impacts of temperature increase and change in precipitation pattern on crop yield and yield quality of barley. Food Chemistry. 2013;**136**:1470-1477

[36] Samarah NH. Effects of drought stress on growth and yield of barley. Agronomy for Sustainable Development. 2005;**25**(1):145-149

[37] Alfonso SU, Brüggemann W. Photosynthetic responses of a C3 and three C4 species of the genus Panicum with different metabolic subtypes to drought stress. Photosynthesis Research. 2012;**112**:175-191

[38] Gomez JM, Jimenz A, Olmas E, Sevilla F. Location and effects of long term NaCl stress on superoxide dismutase and ascorbate peroxidase isoenzymes of pea (*Pisum sativum* cv. Puget) chloroplasts. Journal of Experimental Botany. 2004;**55**:119-130

[39] Shannon MC, Grieve CM. Tolerance of vegetable crops to salinity. Scientia Horticulturae. 1999;**78**:5-38

[40] Salvucci ME, Crafts-Brandner SJ. Relationship between the heat tolerance of photosynthesis and the thermal stability of rubisco activase in plants from contrasting thermal environments. Plant Physiology. 2004;**134**:1460-1470

[41] Polle A, Schützendübel A. Heavy metal signalling in plants: Linking cellular and organismic responses. In: Hirt H, Shinozaki K, editors. Plant Responses to Abiotic Stress. Topics in Current Genetics. Vol. 4. Switzerland: Springer Nature; 2003. pp. 187-215

[42] Jiao S, Hilaire E, Guikema JA. Identification and differential accumulation of two isoforms of the CF1-b subunit under high light stress in *Brassica rapa*. Plant Physiology and Biochemistry. 2004;**42**:883-890

[43] Král M, Dvořák M, Capouchová I. The straw as mulch and compost as a tool for mitigation of drought impacts in the potatoes cultivation. Plant, Soil and Environment. 2019;**65**(11):530-535

[44] Nabati DA, Schmidt RE, Parrish DJ. Alleviation of salinity stress in Kentucky bluegrass by plant growth regulators and iron. Crop Science. 1994;**34**(1):198-202

[45] Shashi M, Mannan M, Islam M, Rahman M. Impact of rice husk biochar on growth, water relations and yield of maize (*Zea mays* L.) under drought condition. The Agriculturists. 2018;**16**: 93-101

[46] Khan S, Wang N, Reid BJ, Freddo A, Cai C. Reduced bioaccumulation of PAHs by *Lactuca sativa* L. grown in contaminated soil amended with sewage sludge and sewage sludge derived biochar. Environmental Pollution. 2013; **175**:64-68

[47] Dong D, Feng Q, Mcgrouther K, Yang M, Wang H, Wu W. Effects of biochar amendment on rice growth and nitrogen retention in a waterlogged paddy field. Journal of Soils and Sediments. 2014;**15**(1):153-162

[48] Jinyang W, Pan X, Liu Y, Zhang X, Xiong Z. Effects of biochar amendment in two soils on greenhouse gas emissions and crop production. Plant and Soil. 2012;**360**(1–2):287-298

[49] Kamara A, Kamara HS, Kamara MS. Effect of rice straw biochar on soil quality and the early growth and biomass yield of two rice varieties. Agricultural Sciences. 2015;**6**:798

[50] Feng J, Cheng R, Qul AA, Yan QG, Li YG, Jian BL, et al. Effects of biochar on sodium ion accumulation, yield and quality of rice in saline-sodic soil of the west of Songnen plain, northeast China. Plant, Soil and Environment. 2018;**64**: 612-618

[51] Ran C, Gulaqa A, Zhu J, Wang X, Zhang S, Geng Y, et al. Benefits of biochar for improving ion contents, cell membrane permeability, leaf water status and yield of rice under saline-sodic paddy field condition. Journal of Plant Growth Regulation. 2020;**39**:370-377

[52] Liu Y, Lu H, Yang S, Wang Y. Impacts of biochar addition on rice yield and soil properties in a cold waterlogged paddy for two crop seasons. Field Crops Research. 2016;**191**:161-167

[53] Haefele S, Konboon Y, Wongboon W, Amarante S, Maarifat A, Pfeiffer E, et al. Effects and fate of biochar from rice residues in rice-based systems. Field Crops Research. 2011;**121**: 430-440

[54] Liu Y, Yang S, Lu H, Wang Y. Effects of biochar on spatial and temporal changes in soil temperature in cold waterlogged rice paddies. Soil and Tillage Research. 2018;**181**:102-109

[55] Huang M, Long F, Jiang LG, Yang SY, Zou YB, Uphoff N. Continuous applications of biochar to rice: Effects on grain yield and yield attributes. Journal of Integrative Agriculture. 2019;**18**: 563-570

[56] Kartika K, Lakitan B, Wijaya A, Kadir S, Widur LI, Siaga E, et al. Effects of particle size and application rate of rice-husk biochar on chemical properties of tropical wetland soil, rice growth and yield. Australian Journal of Crop Science. 2018;**12**:817-826

[57] Fahad S, Hussain S, Saud S, Hassan S, Tanveer M, Ihsan MZ, et al. A combined application of biochar and phosphorus alleviates heat-induced adversities on physiological, agronomical and quality attributes of rice. Plant Physiology and Biochemistry. 2016;**103**: 191-198

[58] He T, Meng J, Chen W, Liu Z, Cao T, Cheng X, et al. Effects of biochar on cadmium accumulation in rice and cadmium fractions of soil: A three-year pot experiment. BioResources. 2017;**12**: 622-642

[59] Bian R, Joseph S, Cui L, Pan G, Li L, Liu X, et al. A three-year experiment confirms continuous immobilization of cadmium and lead in contaminated paddy field with biochar amendment. Journal of Hazardous Materials. 2014; **272**:121-128

[60] Glick BR, Cheng Z, Czarny J, Duan J. Promotion of plant growth by ACC deaminase-producing soil bacteria. European Journal of Plant Pathology. 2007;**119**:329-339

[61] Chi F, Yang P, Han F, Jing Y, Shen S. Proteomic analysis of rice seedlings infected by *Sinorhizobium meliloti* 1021. Proteomics. 2010;**10**:1861-1874

[62] Hussain N, Mujeeb F, Tahir M, Khan GD, Hassan NM, Bari A. Effectiveness of Rhizobium under

salinity stress. Asian Journal of Plant Sciences. 2002;**1**:12-14

[63] Antoun H, Prevost D. Ecology of plant growth promoting rhizobacteria. In: Siddiqui ZA, editor. PGPR: Biocontrol and Biofertilization. Dordrecht: Springer; 2005. pp. 1-38

[64] Yao L, Wu Z, Zheng Y, Kaleem I, Li C. Growth promotion and protection against salt stress by *Pseudomonas putida* Rs-198 on cotton. European Journal of Soil Biology. 2010;**46**:49-54

[65] Egamberdiyeva D. The effect of plant growth promoting bacteria on growth and nutrient uptake of maize in two different soils. Applied Soil Ecology. 2007;**36**:184-189

[66] Alavi P, Starcher MR, Zachow C, Müller H, Berg G. Root-microbe systems: The effect and mode of interaction of stress protecting agent (SPA) *Stenotrophomonas rhizophila* DSM14405T. Frontiers in Plant Science. 2013;**4**:141

[67] Kohler J, Caravaca F. An AM fungus and a PGPR intensify the adverse effects of salinity on the stability of rhizosphere soil aggregates of *Lactuca sativa* Roldan. Soil Biology and Biochemistry. 2010;**42**: 429-434

[68] Gill SS, Khan NA, Tuteja N. Cadmium at high dose perturbs growth, photosynthesis and nitrogen metabolism while at low dose it up regulates sulfur assimilation and antioxidant machinery in garden cress (*Lepidium sativum* L.). Plant Science. 2012;**182**:112-120

[69] Samuel S, Muthukkaruppan SM. Characterization of plant growth promoting rhizobacteria and fungi associated with rice, mangrove and effluent contaminated soil. Current Botany. 2011;**2**:22-25

[70] Baharlouei K, Pazira E, Solhi M. Evaluation of inoculation of plant growth-promoting rhizobacteria on cadmium. Singapore: International Conference on Environmental Science and Technology, IPCBEE. IACSIT Press; 2011; Vol. 6

[71] Bhardwaj D, Ansari MW, Sahoo RK, Tuteja N. Biofertilizers function as key player in sustainable agriculture by improving soil fertility, plant tolerance and crop productivity. Microbial Cell Factories. 2014;**13**:66

[72] Ruiz-Sanchez M, Aroca R, Munoz Y, Polon R, Ruiz-Lozano JM. The arbuscular mycorrhizal symbiosis enhances the photosynthetic efficiency and the antioxidative response of rice plants subjected to drought stress. Journal of Plant Physiology. 2010;**167**: 862-869

[73] Hrkousse O, Simani A, Jadrane I, Aitboulahsen M, Mazri MA, Zouahri A, et al. Role of local biofertilizer in enhancing the oxidative stress defence systems of date palm seedling (*Phoenix dactylifera*) against abiotic stress. Applied and Environmental Soil Science. 2021;**6628544**:1-13

[74] Kohler J, Hernandez JA, Caravaca F, Rold´an A. Plant-growth-promoting rhizobacteria and arbuscular mycorrhizal fungi modify alleviation biochemical mechanisms in waterstressed plants. Functional Plant Biology. 2008;**35**(2):141

[75] Burd GI, Dixon DG, Glick BR. Plant growth-promoting bacteria that decrease heavy metal toxicity in plants. Canadian Journal of Microbiology. 2000;**46**: 237-245

[76] Madhaiyan M, Poonguzhali S, Sa T. Metal tolerating methylotrophic bacteria reduces nickel and cadmium toxicity and

promotes plant growth of tomato (*Lycopersicon esculentum* L.). Chemosphere. 2007;**69**:220-228

[77] Safronova VI, Stepanok VV, Engqvist GL, Alekseyev YV, Belimov AA. Root-associated bacteria containing 1-aminocyclopropane-1-carboxylate deaminase improve growth and nutrient uptake by pea genotypes cultivated in cadmium supplemented soil. Biology and Fertility of Soils. 2006; **42**:267-272

[78] Wani PA, Khan MS, Zaidi A. Effect of metal tolerant plant growth-promoting Rhizobium on the performance of pea grown in metal-amended soil. Archives of Environmental Contamination and Toxicology. 2008;**55**:33-42

[79] Ouledali S, Ennajeh M, Zrig A, Gianinazzi S, Khemira H. Estimating the contribution of arbuscular mycorrhizal fungi to drought tolerance of potted olive trees (Olea europaea). Acta Physiologiae Plantarium. 2018;**40**:1-81

[80] Pavithra D, Yapa N. Arbuscular mycorrhizal fungi inoculation enhances drought stress tolerance of plants. Groundwater for Sustainable Development. 2018;**7**:490-494

[81] Pedranzani H, Rodríguez-Rivera M, Gutiérrez M, Porcel R, Hause B, Ruiz-Lozano JM. Arbuscular mycorrhizal symbiosis regulates physiology and performance of *Digitaria eriantha* plants subjected to abiotic stresses by modulating antioxidant and jasmonate levels. Mycorrhiza. 2016;**26**: 141-152

[82] Rani B. Effect of arbuscular mycorrhiza fungi on biochemical parameters in wheat *Triticum aestivum* L. under drought conditions [Doctoral Dissertation]. Hisar: CCSHAU; 2016

[83] Nelsen CE, Safir GR. Increased drought tolerance of mycorrhizal onion plants caused by improved phosphorus nutrition. Planta. 1982;**154**:407-413

[84] Hajiboland R, Aliasgharzadeh N, Laiegh SF, Poschenrieder C. Colonization with arbuscular mycorrhizal fungi improves salinity tolerance of tomato *Solanum lycopersicum* L. plants. Plant and Soil. 2010;**331**:313-327

[85] Khalloufi M, Martínez-Andújar C, Lachaâl M, Karray-Bouraoui N, Pérez-Alfocea F, Albacete A. The interaction between foliar GA3 application and arbuscular mycorrhizal fungi inoculation improves growth in salinized tomato *Solanum lycopersicum* L. plants by modifying the hormonal balance. Journal of Plant Physiology. 2017;**214**:134-144

[86] Hajiboland R, Dashtebani F, Aliasgharzad N. Physiological responses of halophytic C4 grass, *Aeluropus littoralis* to salinity and arbuscular mycorrhizal fungi colonization. Photosynthetica. 2015;**53**:572-584

[87] Giri B, Kapoor R, Mukerji KG. Improved tolerance of *acacia nilotica*, to salt stress by arbuscular mycorrhiza, *Glomus fasciculatum*, may be partly related to elevated K/Na ratios in root and shoot tissues. Microbial Ecology. 2007;**54**:753-760

[88] Hashem A, Alqarawi AA, Radhakrishnan R, Al-Arjani AF, Aldehaish HA, Egamberdieva D, et al. Arbuscular mycorrhizal fungi regulate the oxidative system, hormones and ionic equilibrium to trigger salt stress tolerance in *Cucumis sativus* L. Saudi Journal of Biological Sciences. 2018;**25**: 1102-1114

[89] Hajiboland R, Joudmand A, Aliasgharzad N, Tolrá R,

Poschenrieder C. Arbuscular mycorrhizal fungi alleviate low-temperature stress and increase freezing resistance as a substitute for acclimation treatment in barley. Crop & Pasture Science. 2019;**70**:218-233

[90] Mathur S, Jajoo A. Arbuscular mycorrhizal fungi protects maize plants from high temperature stress by regulating photosystem II heterogeneity. Industrial Crops and Products. 2020;**143**: 111934

[91] Ma J, Janoušková M, Ye L, Bai LQ, Dong RR, Yan Y, et al. Role of arbuscular mycorrhiza in alleviating the effect of cold on the photosynthesis of cucumber seedlings. Photosynthetica. 2019;**57**: 86-95

[92] Chu XT, Fu JJ, Sun YF, Xu YM, Miao YJ, Xu YF, et al. Effect of arbuscular mycorrhizal fungi inoculation on cold stress-induced oxidative damage in leaves of *Elymus nutans* Griseb. South African Journal of Botany. 2016;**104**: 21-29

[93] Kaldorf M, Kuhn AJ, Schröder WH, Hildebrandt U, Bothe H. Selective element deposits in maize colonized by a heavy metal tolerance conferring arbuscular mycorrhizal fungus. Journal of Plant Physiology. 1999;**154**: 718-728

[94] Kelkar TS, Bhalerao SA. Beneficiary effect of arbuscular mycorrhiza to *Trigonella foenum-graceum* in contaminated soil by heavy metal. Research Journal of Recent Sciences. 2013;**2**:29-32

[95] Jiang QY, Zhuo F, Long SH, Zhao HD, Yang DJ, Ye ZH, et al. Can arbuscular mycorrhizal fungi reduce Cd uptake and alleviate Cd toxicity of Lonicera japonica grown in Cd-added soils? Scientific Reports. 2016;**6**:21805

[96] Lingua G, Franchin C, Todeschini V, Castiglione S, Biondi S, Burlando B, et al. Arbuscular mycorrhizal fungi differentially affect the response to high zinc concentrations of two registered poplar clones. Environmental Pollution. 2008;**153**:137-147

[97] Li XL, Christie P. Changes in soil solution Zn and pH and uptake of Zn by arbuscular mycorrhizal red clover in Zn-contaminated soil. Chemosphere. 2001; **42**:201-207

[98] Miller SP, Sharitz RR. Manipulation of flooding and arbuscular mycorrhiza formation influences growth and nutrition of two semiaquatic grass species. Functional Ecology. 2000;**14**:738-748

[99] Fougnies L, Renciot S, Müller F, Plenchette C, Prin Y, De Faria SM, et al. Arbuscular mycorrhizal colonization and nodulation improve flooding tolerance in Pterocarpus officinalis Jacq. seedlings. Mycorrhiza. 2007;**17**:159-166

[100] Solís-Rodríguez UR, Ramos-Zapata JA, Hernández-Cuevas L, Salinas-Peba L, Guadarrama P. Arbuscular mycorrhizal fungi diversity and distribution in tropical low flooding forest in Mexico. Mycological Progress. 2020;**19**:195-204

[101] Bulgari R, Franzoni G, Ferrante A. Biostimulants application in horticultural crops under abiotic stress conditions. Agronomy. 2019;**9**:306

[102] Pokluda R, Sękara A, Jezdinský A, Kalisz A, Neugebauerová J, Grabowska A. The physiological status and stress biomarker concentration of *Coriandrum sativum* L. plants subjected to chilling are modified by biostimulant application. Biological Agriculture and Horticulture. 2016;**32**:258-268

[103] Marfà O, Cáceres R, Polo J, Ródenas J. Animal protein hydrolysate as

a biostimulant for transplanted strawberry plants subjected to cold stress. Acta Horticulturae. 2009;**842**: 315-318

[104] Polo J, Barroso R, Ródenas J, Azcón-Bieto J, Cáceres R, Marfà O. Porcine hemoglobin hydrolysate as a biostimulant for lettuce plants subjected to conditions of thermal stress. HortTechnology. 2006;**16**:483-487

[105] Botta A. Enhancing plant tolerance to temperature stress with amino acids: An approach to their mode of action. Acta Horticulturae. 2012;**1009**:29-36

[106] Korkmaz A, Korkmaz Y, Demirkiran AR. Enhancing chilling stress tolerance of pepper seedlings by exogenous application of 5-aminolevulinic acid. Environmental and Experimental Botany. 2010;**67**:495-501

[107] Kang SM, Khan AL, Waqas M, You Y-H, Hamayun M, Joo GJ, et al. Gibberellin-producing *Serratia nematodiphila* PEJ1011 ameliorates low temperature stress in *Capsicum annuum* L. European Journal of Soil Biology. 2015;**68**:85-93

[108] Petrozza A, Santaniello A, Summerer S, Di Tommaso G, Di Tommaso D, Paparelli E, et al. Physiological responses to Megafol® treatments in tomato plants under drought stress: A phenomic and molecular approach. Scientia Horticulturae (Amsterdam). 2014;**174**:185-192

[109] Xu C, Leskovar DI. Effects of *A. nodosum* seaweed extracts on spinach growth, physiology and nutrition value under drought stress. Scientia Horticulturae (Amsterdam). 2015;**183**: 39-47

[110] Mayak S, Tirosh T, Glick BR. Plant growth-promoting bacteria that confer resistance to water stress in tomatoes and peppers. Plant Science. 2004;**166**: 525-530

[111] Goñi O, Quille P, O'Connell S. *Ascophyllum nodosum* extract biostimulants and their role in enhancing tolerance to drought stress in tomato plants. Plant Physiology and Biochemistry. 2018;**126**:63-73

[112] Kałuzewicz A, Krzesiński W, Spizewski T, Zaworska A. Effect of biostimulants on several physiological characteristics and chlorophyll content in broccoli under drought stress and re-watering. Notulae Botanicae Horti Agrobotanici Cluj-Napoca. 2017;**45**: 197-202

[113] Petrozza A, Summerer S, Di Tommaso G, Di Tommaso D, Piaggesi A. An evaluation of tomato plant root development and morpho- physiological response treated with VIVA® by image analysis. Acta Horticulturae. 2013;**1009**: 155-160

[114] Heidari M, Golpayegani A. Effects of water stress and inoculation with plant growth promoting rhizobacteria (PGPR) on antioxidant status and photosynthetic pigments in basil (*Ocimum basilicum* L.). Journal of Saudi Society of Agricultural Sciences. 2012;**11**: 57-61

[115] Abd El-Mageed TA, Semida WM, Rady MM. Moringa leaf extract as biostimulant improves water use efficiency, physio-biochemical attributes of squash plants under deficit irrigation. Agricultural Water Management. 2017; **193**:46-54

[116] Asaf S, Hamayun M, Khan AL, Waqas M, Khan MA, Jan R, et al. Salt tolerance of *Glycine max*. L induced by endophytic fungus *Aspergillus flavus* CSH1, via regulating its endogenous

hormones and antioxidative system. Plant Physiology and Biochemistry. 2018;**128**:13-23

[117] Barnawal D, Bharti N, Pandey SS, Pandey A, Chanotiya CS, Kalra A. Plant growth-promoting rhizobacteria enhance wheat salt and drought stress tolerance by altering endogenous phytohormone levels and TaCTR1/TaDREB2 expression. Physiologia Plantarum. 2017;**161**:502-514

[118] Kang SM, Khan MA, Hamyun M, Kim LR, Kwon EH, Kang YS, et al. Phosphate-solubilizing *Enterobacter ludwigii* AFFR02 and *Bacillus megaterium* Mj1212 rescues alfalfa's growth under post-drought stress. Agriculture. 2021;**11**:485

[119] Niu X, Song L, Xiao Y, Ge W. Drought-tolerant plant growth-promoting rhizobacteria associated with foxtail millet in a semi-arid agroecosystem and their potential in alleviating drought stress. Frontiers in Microbiology. 2018;**8**:2580

[120] Batool T, Ali S, Seleiman MF, Naveed NH, Ali A, Ahmed K, et al. Plant growth promoting rhizobacteria alleviates drought stress in potato in response to suppressive oxidative stress and antioxidant enzymes activities. Scientific Reports. 2020;**2020**(10):1-19

[121] Begum N, Wang L, Ahmad H, Akhtar K, Roy R, Khan MI, et al. Co-inoculation of arbuscular mycorrhizal fungi and the plant growth-promoting rhizobacteria improve growth and photosynthesis in tobacco under drought stress by up-regulating antioxidant and mineral nutrition metabolism. Microbial Ecology. 2022;**83**:971-988

[122] Chandra D, Srivastava R, Glick BR, Sharma AK. Drought-tolerant Pseudomonas spp. improve the growth performance of finger millet (*Eleusine coracana* (L.) Gaertn.) under non-stressed and drought-stressed conditions. Pedosphere. 2018;**28**(2):227-240

[123] Yasmin H, Nosheen A, Naz R, Bano A, Keyani R. L-tryptophan-assisted PGPR-mediated induction of drought tolerance in maize (*Zea mays* L). Journal of Plant Interactions. 2017;**12**(1):567-578

[124] Abdel Megeed TM, Gharib HS, Hafez EM, El-Sayed A. Effect of some plant growth regulators and biostimulants on the productivity of Sakha108 rice plant (*Oryza sativa* L.) under different water stress conditions. Applied Ecology and Environmental Research. 2021;**19**(4):2859-2878

[125] Francesca S, Cirillo V, Raimondi G, Maggio A, Barone A, Rigano MM. A novel protein hydrolysate-based biostimulant improves tomato performances under drought stress. Plants. 2021;**10**:1-13

[126] Ali S, Khan MA, Kim W-CJABC. *Pseudomonas veronii* KJ mitigates flood stress-associated damage in *Sesamum indicum* L. Applied Biological Chemistry. 2018;**2018**(61):575-585

[127] Rauf M, Awais M, Ud-Din A, Ali K, Gul H, Rahman MM, et al. Molecular mechanisms of the 1-aminocyclopropane-1-carboxylic acid (ACC) deaminase producing *Trichoderma asperellum* MAP1 in enhancing wheat tolerance to waterlogging stress. Frontiers in Plant Science. 2021;**11**:2213

[128] El-Bassiony AM, Ghoname AA, El-Awadi ME, Fawzy ZF, Gruda N. Ameliorative Effects of brassinosteroids on growth and productivity of snap beans grown under high temperature. Gesunde Pflanzen. 2012;**64**:175-182

[129] Nahar K, Hasanuzzaman M, Alam MM, Fujita M. Exogenous glutathione confers high temperature stress tolerance in mung bean (*Vigna radiata* L.) by modulating antioxidant defense and methylglyoxal detoxification system. Environmental and Experimental Botany. 2015;**112**:44-54

[130] Kaushal N, Gupta K, Bhandhari K, Kumar S, Thakur P, Nayyar H. Proline induces heat tolerance in chickpea (Cicer arietinum L.) plants by protecting vital enzymes of carbon and antioxidative metabolism. Physiology and Molecular Biology of Plants. 2011;**17**:203-213

[131] Kumar S, Kaushal N, Nayyar H, Gaur P. Abscisic acid induces heat tolerance in chickpea (*Cicer arietinum* L.) seedlings by facilitated accumulation of osmoprotectants. Acta Physiologiae Plantarum. 2012;**34**:1651-1658

[132] Khan MA, Asaf S, Khan AL, Jan R, Kang SM, Kim KM, et al. Extending thermotolerance to tomato seedlings by inoculation with SA1 isolate of *Bacillus cereus* and comparison with exogenous humic acid application. PLoS One. 2020; **15**(4):e0232228

[133] Kang SM, Khan AL, Waqas M, Asaf S, Lee KE, Park YG, et al. Integrated phytohormone production by the plant growth-promoting rhizobacterium *Bacillus tequilensis* SSB07 induced thermotolerance in soybean. Journal of Plant Interactions. 2019;**14**:416-423

[134] Quintero-Calderón EH, Sánchez-Reinoso AD, Chávez-Arias CC, Garces-Varon G, Restrepo-Díaz H. Rice seedlings showed a higher heat tolerance through the foliar application of biostimulants. Notulae Botanicae Horti Agrobotanici Cluj-Napoca. 2021;**49**:1

[135] Niu C, Wang G, Sui J, Liu G, Ma F, Bao Z. Biostimulants alleviate temperature stress in tomato seedlings. Scientia Horticulturae. 2021;**2022**(293): 110712

[136] Spinelli F, Fiori G, Noferini M, Sprocatti M, Costa G. A novel type of seaweed extract as a natural alternative to the use of iron chelates in strawberry production. Scientia Horticulturae (Amsterdam). 2010;**125**:263-269

[137] Cerdán M, Sánchez-Sánchez A, Jordá JD, Juárez M, Sánchez-Andreu J. Effect of commercial amino acids on iron nutrition of tomato plants grown under lime-induced iron deficiency. Journal of Plant Nutrition and Soil Science. 2013; **176**:859-866

[138] Papenfus HB, Kulkarni MG, Stirk WA, Finnie JF, Van Staden J. Effect of a commercial seaweed extract (Kelpak®) and polyamines on nutrient-deprived (N, P and K) okra seedlings. Scientia Horticulturae (Amsterdam). 2013;**151**:142-146

[139] Anjum K, Ahmed M, Baber JK, Alizai MA, Ahmed N, Tareen MH. Response of garlic bulb yield to bio-stimulant (Bio-cozyme) under calcareous soil. Life Sciences: An International Journal. 2014;**8**:3058-3062

[140] Barassi CA, Ayrault G, Creus CM, Sueldo RJ, Sobrero MT. Seed inoculation with *Azospirillum* mitigates NaCl effects on lettuce. Scientific Horticulturae (Amsterdam). 2006;**109**:8-14

[141] Del Amor FM, Cuadra-Crespo P. Plant growth-promoting bacteria as a tool to improve salinity tolerance in sweet pepper. Functional Plant Biology. 2012;**39**:82-90

[142] Del Pilar CM, Berrido SI, Ligero F, Lluch C. Rhizobium strain effects on the growth and nitrogen assimilation in Pisum sativum and *Vicia faba* plant

growth under salt stress. Journal of Plant Physiology. 1999;**154**:127-131

[143] Yildirim E, Taylor AG, Spittler TD. Ameliorative effects of biological treatments on growth of squash plants under salt stress. Scientific Horticulturae (Amsterdam). 2006;**111**:1-6

[144] Aydin A, Kant C, Turan M. Humic acid application alleviate salinity stress of bean (*Phaseolus vulgaris* L.) plants decreasing membrane leakage. African Journal of Agricultural Research. 2012;**7**: 1073-1086

[145] Ross R, Holden D. Commercial extracts of the brown seaweed *Ascophyllum nodosum* enhance growth and yield of strawberries. HortScience. 2010;**45**:S141-S141

[146] Guinan KJ, Sujeeth N, Copeland RB, Jones PW, O'Brien NM, Sharma HSS, et al. Discrete roles for extracts of *Ascophyllum nodosum* in enhancing plant growth and tolerance to abiotic and biotic stresses. Acta Horticulturae. 2013;**1009**:127-135

[147] Lucini L, Rouphael Y, Cardarelli M, Canaguier R, Kumar P, Colla G. The effect of a plant-derived biostimulant on metabolic profiling and crop performance of lettuce grown under saline conditions. Scientific Horticulturae (Amsterdam). 2015;**182**:124-133

[148] Mayak S, Tirosh T, Glick BR. Plant growth-promoting bacteria confer resistance in tomato plants to salt stress. Plant Physiology and Biochemistry. 2004;**42**:565-572

[149] Demir K, Günes A, Inal A, Alpaslan M. Effects of humic acids on the yield and mineral nutrition of cucumber (*Cucumis Sativus*, L.) grown with different salinity levels. Acta Horticulturae. 1999;**492**:95-104

[150] Rady MM, Desoky ESM, Elrys AS, Boghdady MS. Can licorice root extract be used as an effective natural biostimulant for salt-stressed common bean plants? South African Journal of Botany. 2019;**121**:294-305

[151] Semida WM, Rady MM. Presoaking application of propolis and maize grain extracts alleviates salinity stress in common bean (*Phaseolus vulgaris* L.). Scientific Horticulturae (Amsterdam). 2014;**168**:210-217

[152] Rady MM, Varma B, Howladar SM. Common bean (*Phaseolus vulgaris* L.) seedlings overcome NaCl stress as a result of presoaking in *Moringa oleifera* leaf extract. Scientific Horticulturae (Amsterdam). 2013;**162**:63-70

[153] Abdel Latef AAH, Srivastava AK, Saber H, Alwaleed EA, Tran LSP. *Sargassum muticum* and *Jania rubens* regulate amino acid metabolism to improve growth and alleviate salinity in chickpea. Scientific Reports. 2017;**7**:1-12

[154] Arroussi HE, Benhima R, Elbaouchi A, Sijilmassi B, Mernissi NE, Aafsar A, et al. *Dunaliella salina* exopolysaccharides: A promising biostimulant for salt stress tolerance in tomato (*Solanum lycopersicum*). Journal of Applied Phycology. 2018;**30**: 2929-2941

[155] Semida WM, Abd El-Mageed TA, Hemida K, Rady MM. Natural bee-honey based biostimulants confer salt tolerance in onion via modulation of the antioxidant defence system. The Journal of Horticultural Science and Biotechnology. 2019;**94**:1-11

[156] Mesut Çimrin K, Türkmen Ö, Turan M, Tuncer B. Phosphorus and humic acid application alleviate salinity stress of pepper seedling. African Journal of Biotechnology. 2010;**9**:5845-5851

[157] Sapre S, Gontia-Mishra I, Tiwari S. Plant growth-promoting rhizobacteria ameliorates salinity stress in pea (*Pisum sativum*). Journal of Plant Growth Regulation. 2022;**41**:647-656

[158] Mellidou I, Ainalidou A, Papadopoulou A, Leontidou K, Genitsaris S, Karagiannis E, et al. Comparative transcriptomics and metabolomics reveal an intricate priming mechanism involved in PGPR-mediated salt tolerance in tomato. Frontiers in Plant Science. 2021;**12**:713984

[159] Alexander A, Singh VK, Mishra A. Halotolerant PGPR *Stenotrophomonas maltophilia* BJ01 induces salt tolerance by modulating physiology and biochemical activities of *Arachis hypogaea*. Frontiers in Microbiology. 2020;**11**:1-12

[160] Gupta S, Pandey S. ACC deaminase producing bacteria with multifarious plant growth promoting traits alleviates salinity stress in french bean (*Phaseolus vulgaris*) plants. Frontiers in Microbiology. 2019;**10**:1506

[161] Khan MA, Asaf S, Khan AL, Adhikari A, Jan R, Ali S, et al. Halotolerant rhizobacterial strains mitigate the adverse effects of Nacl stress in soybean seedlings. BioMed Research International. 2019; **2019**:9530963

[162] Ikram M, Ali N, Jan G, Iqbal A, Hamayun M, Jan FG, et al. *Trichoderma reesei* improved the nutrition status of wheat crop under salt stress. Journal of Plant Interactions. 2019;**14**(1):590-602

[163] Zhang S, Gan Y, Xu B. Application of plant-growth-promoting fungi *Trichoderma longibrachiatum* T6 enhances tolerance of wheat to salt stress through improvement of antioxidative defense system and gene expression. Frontiers in Plant Science. 2016;**7**:1405

[164] Hamayun M, Hussain A, Khan SA, Kim HY, Khan AL, Waqas M, et al. Gibberellins producing endophytic fungus *Porostereum spadiceum* AGH786 rescues growth of salt affected soybean. Frontiers in Microbiology. 2017;**8**:686

[165] Souza AC, Zandonadi DB, Santos MP, Canellas NOA, de Paula SC, da Silva Irineu LES, et al. Acclimation with humic acids enhances maize and tomato tolerance to salinity. Chemical and Biological Technologies in Agriculture. 2021;**8**(1):1-13

[166] Khan MS, Pandey MK, Hemalatha S. Comparative studies on the role of organic biostimulant in resistant and susceptible cultivars of rice grown under saline stress—organic biostimulant alleviate saline stress in tolerant and susceptible cultivars of rice. Journal of Crop Science and Biotechnology. 2018;**21**(5):459-467

[167] El Boukhari MELM, Barakate M, Bouhia Y, Lyamlouli K. Trends in seaweed extract based biostimulants: manufacturing process and beneficial effect on soil-plant systems. Plants. 2020;**9**(3):359

[168] Nawaz A, Shahbaz M, Asadullah M, Imran A, Marghoob MU, Imtiaz M, et al. Potential of salt tolerant PGPR in growth and yield augmentation of wheat (*Triticum aestivum* L) under saline conditions. Frontiers in Microbiology. 2020;**11**:2019

[169] Jan R, Khan MA, Asaf S, Lubna LIJ, Kim KM. Metal resistant endophytic bacteria reduces cadmium, nickel toxicity, and enhances expression of metal stress related genes with improved growth of oryza sativa, via regulating its antioxidant machinery and endogenous hormones. Plants (Basel). 2019;**8**(10):363

[170] Kang SM, Asaf S, Khan AL, Lubna KA, Mun BG, Khan MA, et al.

Complete genome sequence of *Pseudomonas psychrotolerans* CS51, a plant growth-promoting bacterium, under heavy metal stress conditions. Microorganisms. 2020;**8**(3):382

[171] Sahile AA, Khan MA, Hamayun M, Imran M, Kang SM, Lee IJ. Novel *Bacillus cereus* Strain, ALT1, enhance growth and strengthens the antioxidant system of soybean under cadmium stress. Agronomy. 2021;**11**:404

[172] Abdallah EF, Abeer H, Alqarawi AA, Hend AA. Alleviation of adverse impact of cadmium stress in sunflower (*Helianthus annuus* L.) by arbuscular mycorrhizal fungi. Pakistan Journal of Botany. 2015;**47**(2):785-795

Chapter 3

Artificial Photosynthesis as an Alternative Source of Renewable Energy: Potential and Limitations

Renu Kathpalia and Anita Kamra Verma

Abstract

Artificial photosynthesis system (APS) uses biomimetic systems to duplicate the process of natural photosynthesis that utilizes copious resources of water, carbon dioxide and sunlight to produce oxygen and energy-rich compounds and has potential to be an alternative source of renewable energy. APS like natural photosynthesis includes the splitting of water into oxygen and hydrogen, and the reduction of carbon dioxide into various hydrocarbons such as formic acid (HCOOH), methane (CH_4) and carbon monoxide (CO), or even pure hydrogen fuel. These processes are accomplished by a handful of device designs, including photoelectrochemical cells or photovoltaic-coupled electrolyzers which are driven by energy extracted from sunlight photons as well as suitable catalysts. Researchers are trying to combine advantageous components from both natural photosynthesis and artificial photosynthesis to create a semi-artificial photosynthesis system, involving the incorporation of enzymes or even whole-cell into synthetic devices. However, there are several limitations to the advancement of this field which are mainly centered on the inability to establish a system that is cost-effective, long-term durable and has the highest efficiency. Artificial photosynthesis devices can also function as atmospheric cleansers by extracting the excess amount of carbon dioxide and releasing back oxygen into the environment. Although there is still a long way to go to empower society with energy supplied through artificial photosynthesis, at the same time it is both desirable and necessary. To date, the efforts to commercialize APS have been fruitful, and it will soon be a viable alternative fuel source.

Keywords: artificial photosynthesis system (APS), biomimicry, photocatalyst, photosynthesis, photons, water oxidation

1. Introduction

In this technological era, it has become mandatory to safeguard our natural resources and search for renewable sources of energy that can reduce the use of conventional fossil fuels. The usage of fossil fuel emits large quantities of carbon dioxide, sulfur dioxide and oxide particles as well as depletes natural resources. The International Panel on climate change recommended the urgent need to decrease carbon dioxide emissions to zero on a global scale. To combat this grieving situation the development of sustainable and carbon-neutral energy technologies is the most compelling challenge faced by the entire humanity. Over the years, scientists have explored numerous alternatives that could possibly reduce our dependence on fossil fuels. Recently, efforts are centered toward the development of high-tech energy-generating systems inspired by nature itself. Nearly all natural resources are either depleting or getting contaminated except the solar energy which on conversion or utilization is a promising solution for energy-related problems.

One such phenomenon or reaction that takes place in plants, algae and photosynthetic bacteria to produce energy for themselves and provide energy to other organisms is photosynthesis. To mimic photosynthesis, the concept of artificial photosynthesis was introduced by Giacomo Ciamician way back in the year 1912 in a science paper entitled "The photochemistry of the future" [1]. He visualized and insisted on the use of technologies that can eliminate complete dependence on fossil fuels. Solar power stations could meet the challenges of sustainable energy more over the cost is lower than that of nuclear and thermal plants but the main bottleneck is the lack of efficient storage solution [2]. Different studies established systems to carry out photosynthesis similar to that of solar panels and convert it into electricity for direct application, as it was not possible to store energy for later use. So far, researchers have not been able to devise solar-driven catalysts for water oxidation and fuel production that are vigorous and use this abundant earth element.

Artificial photosynthesis system (APS) imitates the fundamental process of photosynthesis taking place in organisms for our societal needs. APS captures as well as stores solar energy in the form of fuel rather than glucose and is able to meet both the challenges of being carbon-negative and a source of solar fuel (**Figure 1**). Artificial photosynthesis devices involving semiconductors can absorb solar energy and store it by converting in the form of chemical energy which can be used later. Many

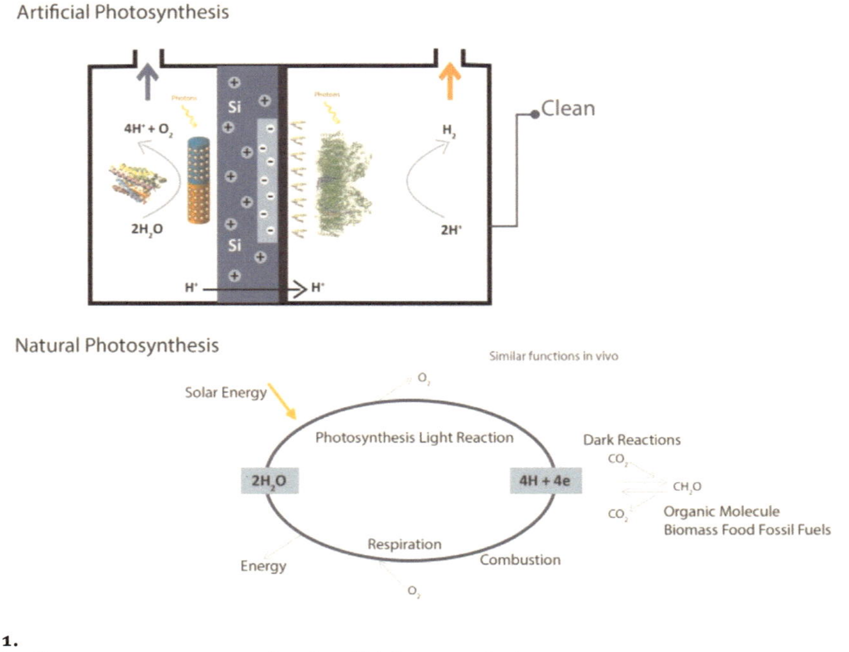

Figure 1.
An outline diagram to compare natural and artificial photosynthesis.

advancements have been achieved where artificial reaction center involving the movement of electrons (injected from a dye) into the conduction band of nanoparticles (such as titanium dioxide) on to electrode coupled with catalyst (such as platinum) or hydrogenase enzymes producing hydrogen gas. This system involves solar energy to split the water molecule into oxygen and hydrogen fuel; however, the efficiency is quite low and requires external electrical potential. The energy stored in this way is cheap and dense as compared to expensive battery storage [3]. Besides these APS is more environmentally attractive than solar panels as it absorbs excess carbon dioxide from the environment and releases oxygen back into the environment and can revolutionize the world of solar power [4]. This chapter includes a comprehensive view of the artificial photosynthesis system, its limitation, challenges and the future scope of APS as an alternate source of energy.

2. Photosynthesis: nature's marvel

Green plants, algae and photosynthetic bacteria are photosynthesizing for more than a billion years without any significant change. Photosynthesis is an integral part of simulation models to safeguard the future of our planet. The electron transport system of photosynthesis transports electrons faster than the photons of solar energy reaching light-harvesting complex present in the chloroplast membrane. The light reaction takes place in three distinct protein complexes which are an integral part of the chloroplast membrane of higher plants viz., two light-harvesting complexes or photosystem (LHC I/PS I and LHCII/PSII) or antennae and cytochrome b_6f connected by different electron carriers. Both the photosystems or light-harvesting complexes transfer photons to the reaction centers (RCs) in the form of resonance energy and create charge differences across the membrane. A strong oxidant complex commonly referred to as oxygen-evolving complex (OEC), is present on the donor side of PSII which does photolysis of water into molecular oxygen, protons and electrons. This electron moves to the plastoquinone pool to cyt b_6f followed by plastocyanin to PSI and finally via ferrodoxin reduces $NADP^+$ to NADPH. The electron movement is coupled with proton pumping from one side of the membrane to the other side of the membrane creating a potential and pH difference across the membrane ultimately leading to the formation of ATP [5]. This assimilatory power trapped in the light reaction is utilized for the unique process of fixation of atmospheric carbon dioxide by the enzyme Rubisco, the most abundant protein on this earth. The enzyme Rubisco has a very slow catalytic rate (1–3 cycles per second) creating a major bottleneck in increasing photosynthetic efficiency. The unique feature of this fundamental process of life is the structural features of RCs complexes constituting a network of molecular cofactors held by the membrane at an appropriate distance and orientation to capture maximum light and perform the movement of electrons (**Figure 2**). This mosaic structure is such a unique nanoscale complex working perfectly for billions of years and providing energy to all the heterotrophs residing on this planet. To exactly mimic the complete process of photosynthesis is definitely an undaunted task.

3. Artificial photosynthesis

The concept of artificial photosynthesis was given by an Italian chemist in 1912 but no remarkable research has been done till 1972 when Kenichi Honda and his student

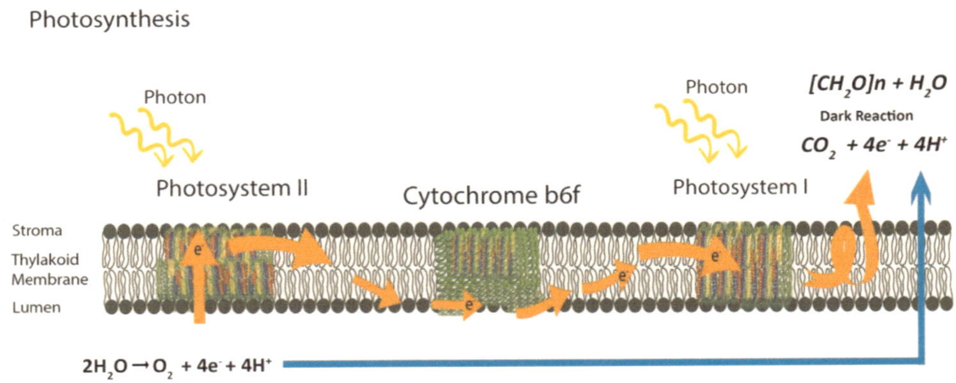

Figure 2.
The diagrammatic representation of different complexes involved in electron transfer in the Z scheme present on the membrane of thylakoid in the chloroplast. The photosystems PSI and PS II and intermediate carrier of the electron, cytochrome b_6f, sand witched between them is the key structural organization of RCs. The complete structure is present in the membrane protein scaffold at an appropriate distance to facilitate electron transfer and convert solar energy to chemical energy.

Akira Fujishima for the first time reported the successful designing of water splitting device powered by light [6], which was named as "Honda-Fujishima effect". The device includes a photoelectrochemical cell comprised of a photoanode made of TiO_2 and a black cathode made up of platinum (Pt) black cathode. Both anode and cathode were completely submerged during exposure to light, an electron was released from TiO_2 leading to the formation of an "electron hole", or positive charge on the Ti atom which was filled by an electron released from a water molecule, oxidizing the water to produce oxygen. The released electron was donated to a proton derived from water, thus reducing the proton to form hydrogen. The light of wavelength more than

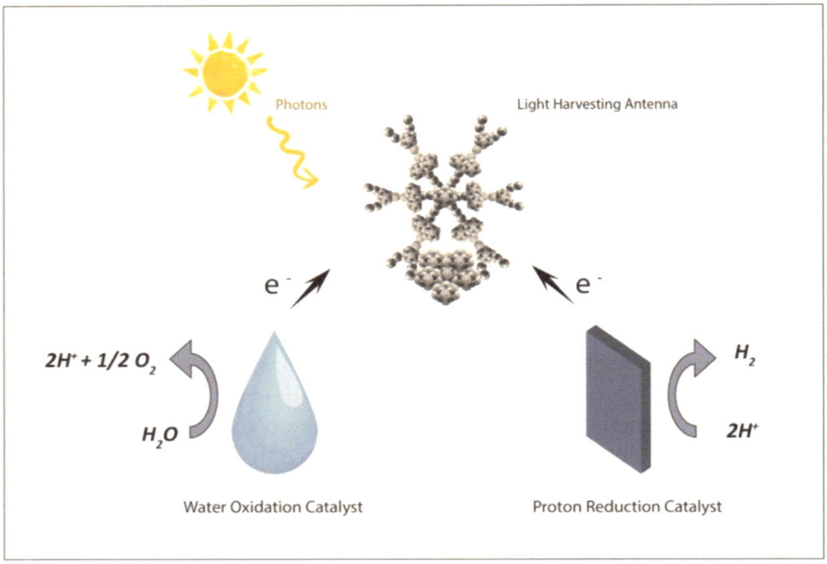

Figure 3.
The diagram representing the fundamental process of artificial photosynthesis systems having an efficient light absorber to trap sunlight and efficient water oxidation and proton reduction catalysts.

400nm was given to the photoanode and the device was able to generate oxygen at the anode while hydrogen was generated at the cathode by photolysis of water and ultimately releasing oxygen and hydrogen [7] as summarized in **Figure 3**. The major bottleneck of using the molecule TiO_2 was that it only absorbed ultraviolet wavelengths and is inactive in visible light wavelength [8].

In 1978, M. Halmann used a p-type phosphide semiconductor made of gallium phosphide as the photocathode suspended in an aqueous solution and achieved the reduction of carbon dioxide into hydrocarbons such as formic acid (HCOOH), formaldehyde (CH_2O) and methanol (CH_3OH) [9]. In 1983, William Ayers designed and patented the first visible light water-splitting device named as "artificial leaf" made up of silicon with a nafion membrane for ion transport above the cell. The entire structure is immersed in water and when illuminated results in the release of oxygen from the back of the metal surface, while hydrogen evolved on the silicon surface [10].

4. Artificial photosynthesis: principle

As mentioned above the three principle steps involved in artificial photosynthesis, similar to natural photosynthesis, are the absorption of light causing excitation, charge generation followed by charge separation and finally chemical conversion leading to the production of fuel.

4.1 Light absorption

In natural photosynthesis, chlorophylls and carotenoids are arranged in antennae to capture the maximum red and blue wavelength of light and bring about the excitation of electrons. These pigment molecules can absorb only a limited range of wavelengths ranging from about 400 to 700 nm which makes less than 50% of the sunlight that reaches earth [11]. The first major task is to design photosensitizers that can optimally use photons on exposure and are capable of aggregating light energy. In addition, the materials used should be capable of absorbing a wider wavelength of the solar spectrum so as to extract maximum energy falling on the earth.

Inspired by the "Honda-Fujishima effect", many materials such as TiO_2 photoanode semiconductors like silicon, metal oxides such as ZnO, Fe_2O_3 and $BiVO_4$, metal nitrides such as Ta_3N_5, metal phosphides such as GaP, metal oxynitrides such as TaON etc., have been tried by a different group of researchers [12]. Silicon, an abundant and cheap source, can absorb a wider spectrum of light. Another semiconductor Gallium Nitride has been used to produce formic acid and ethanol using thin film technology [13].

4.2 Lysis of water

Water oxidation is a thermodynamically uphill process and requires free energy of $\Delta G \approx 237$ kJ mol^{-1} and potential E0 ≈ 1.23 V to transfer $4H^+$ and $4e^-$. In natural photosynthesis, water splitting is achieved through the involvement of an oxygen-evolving complex (OEC) which has manganese (Mn), a tetrameric high valent oxo species, that catalyzes oxygen-oxygen bond formation. Semiconductor nanowires are used to absorb light resulting in the oxidation of water producing oxygen, proton and electrons. The electrons move toward other ends while protons move through a proton-conducting membrane made up of Nafion, ultimately reduced to hydrogen.

Thus, photolysis of water in APS is achieved through the combination of two different customized systems for their respective purpose [12].

The redox equations involved in water splitting are as follows:

$$\text{Oxidation reaction}: 2H_2O \rightarrow 4e^- + 4H^+ + O_2 \quad (1)$$

$$\text{Reduction reaction}: 4H^+ + 4e^- \rightarrow 2H_2 \quad (2)$$

$$\text{Redox reaction}: 2H_2O \rightarrow 2H_2 + O_2 \quad (3)$$

The splitting of water needs an energy of approximately 2.5 V, hence a catalyst is essential to absorb photons of sunlight and set off the reactions [14]. The bioinspired approach of using manganese as a catalyst resulted in instability due to short-term and inefficient function [15]. In comparison, cobalt oxide (CoO) found to be stable, efficient and is easily available [16].

Different materials tested show efficiency in some steps and inefficiency in other steps leading to the use of coupled materials, which are customized for their respective reactivity. Recently, molecular water-oxidation catalysts are designed for splitting water and evolution of oxygen. These catalysts generally comprise a metal complex with wide open coordination sites as well as an electronic structure to stabilize a metal-hydride intermediate. The most common materials used are noble metals such as rhodium and platinum-based complexes. To develop catalysts from earth-abundant metals such as cobalt, iron, molybdenum, and nickel have also been tried by various scientists. The most stable and efficient catalyst out of many tried and tested is Nickel complexes. Ruthenium and iridium-based catalysts showed good reactivity and stability but are scarce and expensive [17]. The transition metal family such as copper, nickel and iron-based were tested in order to improve the catalytic function. Cobalt and Zirconium heterobimetallic on porous silica separated by nanotube separation membranes was also used by researchers [18]. However, the search and optimization of materials to be used in APS are still under investigation.

4.3 Reduction of carbon dioxide

Carbon dioxide is a linear, highly stable molecule having very low electron affinity. Conversion of carbon dioxide is an uphill reaction and requires a nucleophilic attack on the carbon atom, the dissociation bond energy of C=O is ~750 kJ/mol. The carbon atom has the highest valence due to which different fuels can be created in addition to the production of oxygen and hydrogen derived from water. The different fuel compounds that can be generated are formic acid (HCOOH), methanol (CH_3OH), carbon monoxide (CO) and methane (CH_4) as given in these equations:

$$CO_2 + 2H^+ + 2e^- \rightarrow HCOOH \quad (4)$$

$$CO_2 + 2H^+ + 2e^- \rightarrow CO + H_2O \quad (5)$$

$$CO_2 + 6H^+ + 6e^- \rightarrow CH_3OH + H_2O \quad (6)$$

$$CO_2 + 8H^+ + 8e^- \rightarrow CH_4 + 2H_2O \quad (7)$$

The generation of these liquid hydrocarbons has added advantage of being easily integrated into energy infrastructure. Nevertheless, the greatest scientific challenge due to carbon's multi-electron nature imposes additional complexity [19]. The exact mechanism is not clearly established but it is predicted that the reduction of carbon dioxide is similar to water splitting during the process of photosynthesis. In APS, carbon dioxide interacts with catalyst along with the transfer of electrons resulting in a reduction reaction followed by proton coupling creating an electron hole in the photosensitizer which is filled by an electron leading to the end of the process [20]. A co-catalysts and photoelectrode material break carbon and oxygen bonds and form carbon and hydrogen bond i.e., C—H bond. The choice of catalyst is very critical in APS which includes low cost and durability. Researchers have tried various combinations of complexes viz., rhenium-based, cobalt-based, nickel-based, iron-based and zinc-based complexes [21]. With the use of innovative porous materials adsorption of intermediates at the surface of the electrode is significantly enhanced. In another trial, co-catalyst such as copper, silver and gold were found to be very effective in enhancing conserving efficiency. However, an ideal catalyst is yet to design which can improve the performance of chemical conversion (**Figure 4**).

5. Natural photosynthesis vs artificial photosynthesis

Photosynthesis involves two reactions viz, light-dependent (light reaction) harnessing light photons and converting it in assimilatory powers in the form NADPH (nicotinamide adenine dinucleotide phosphate) and ATP (adenosine triphosphate) which are utilized to fix carbon in light-independent reaction (dark reaction). The light reaction involves two photosystems, viz., PSI and PSII (photosystem I and II) consisting of light-absorbing pigments in the form of antennae system involving chlorophylls and carotenoids and have a reaction center having chlorophyll a molecule. The two photosystems are present in the thylakoid membrane along with other electron carriers, absorb light photons and transfer the energy to the reaction center through resonance energy transfer and release an electron from chlorophyll molecule

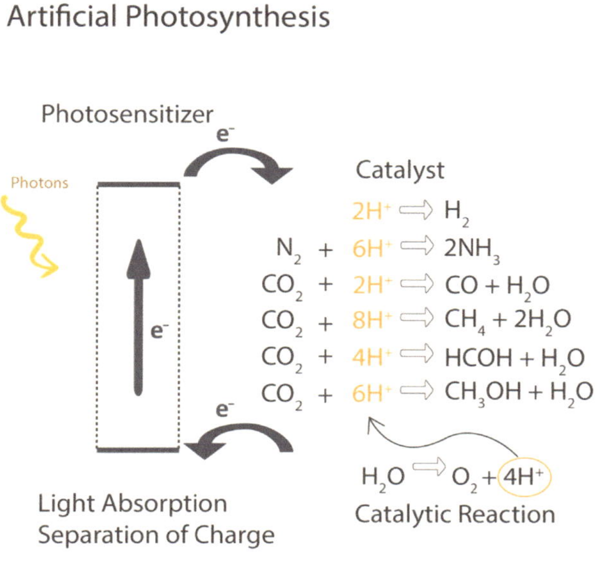

Figure 4.
The formation of different end products due to the reduction of carbon dioxide catalyzed by different catalysts in artificial photosynthesis.

from the reaction center. This electron moves through a series of electron carriers and finally reduces NADP into NADPH and the formation of ATP. The removal of electrons from chlorophyll results in the formation of "electron hole" in the chlorophyll pigment which is filled by photolysis of water by oxygen-evolving complex present next to PSII releasing an electron and oxygen as a waste product for the plants and the most valuable by-product for the survival of life on this biosphere. This light-dependent reaction also involves a complex array of enzymes such as the photosystems themselves, as well as hydrogenases which interact with hydrogen derived from water molecules, is known as photophosphorylation; a unique system of generation of chemical energy present only in photosynthesizing organisms [4, 5].

The assimilatory powers being generated during light reaction then bring about a reduction of carbon through the light-independent reaction or dark reaction involving the Calvin cycle. The dark reaction, discovered by Malvin Calvin, occurs in the stroma of chloroplast through a series of reactions catalyzed by different enzymes to transform carbon dioxide from the atmosphere into carbohydrates. The carbon dioxide is absorbed by the stomata present on the leaves and ultimately converted into carbohydrates, the overall reaction of photosynthesis is

$$6\,H_2O + 6\,CO_2 \frac{\text{Sunlight}}{\text{Chlorophyll}} \rightarrow C_6H_{12}O_6 + 6O_2 \tag{8}$$

Thus, photosynthesis is ingeniously designed through series evolution to provide autotrophic organisms to produce their own food by converting solar energy into chemical energy stored in the energy-rich bonds of a carbohydrate.

Three major components of natural photosynthesis that need to be imitated by artificial photosynthesis are (i) light capture and electron transport (ii) water splitting (into hydrogen and oxygen) and (iii) reduction of carbon dioxide. Two types of fuel are being generated through APS viz., hydrocarbons (methanol and formic acid) and pure hydrogen. Hydrogen is a clean option generated via APS and can be used directly as liquid fuel, consumed in the fuel cell, thermal processes, electrolysis, biological processes and other application to substitute fossil fuel [2, 22].

6. Solar harvesting devices vs artificial photosynthesis

Various efforts have been made to design the devices with the aim to convert solar energy into chemical energy which is stored in chemical bonds. The two most powerful devices designed by researchers are the photoelectrochemical cell and photovoltaic-coupled electrolyzer. Both the systems are designed on certain similarities and differences as well as having advantages and limitations which are discussed here.

6.1 Photoelectrochemical cell

The photoelectrochemical cells (PEC) consist of two electrodes, a photoanode and a photocathode immersed in an electrolyte and an external wire [19, 22]. One of the variants designed comprises a single electrode evolving mixture of oxygen and hydrogen which entails contamination which is an additional step of separating hydrogen and oxygen. The reduction of carbon dioxide by PEC is the most efficient method involving the amalgamation of photo and electrocatalysis, an addition of biocatalysts enhanced its efficiency further. The photoelectrodes are constructed using either a molecular electrode or a light-absorbing semiconductor [23]. At the photoanode, water splits to get oxidized to form oxygen while it is reduced to form hydrogen at the photocathode. On exposure to solar energy excitation at the photoanode releases an electron which is donated to water leading to the reduction of hydrogen molecules at the photocathode. The "electron hole" created at the photoanode is filled by an electron donated by water molecule oxidation ultimately leading to the production of oxygen (**Figure 5**).

The material of the photoanode has a tremendous effect on water-splitting efficiency, stability and absorption of visible light in an aqueous solution (**Figure 6**). A membrane composed of Nafion, having proton-conductive and separation properties, has also been tried in some photoelectrochemical cells. Even semiconductors have also been tried but found to be having low efficiency, instability and only narrow visible light can be adsorbed. Therefore, the criteria of selection shifted toward carbon-based, transition metals, and nanostructured photoanodes e.g., carbon-based materials, graphene, carbon nanotubes, carbon dots and carbon quantum dots all demonstrated stability as well as photocurrent generation. In addition, transition metals such as germanium-doped hematite, cadmium sulfide, zinc oxide, copper sulfide and molybdenum exhibited high performance due to their high electrical conduction and electrochemical stability. Nanomaterials such as nanowires, nanotapers and nanorods also exhibited excellent performance in terms of high hydrogen evolution at photocathode [24]. A serially coupled PSII and PSI along with an Au electrode has been developed recently where semi-artificial PECs produced photocurrent and imitated Z scheme of natural photosynthesis, interfaced by two types of redox osmium complex hydrogel for the transfer of electrons [23].

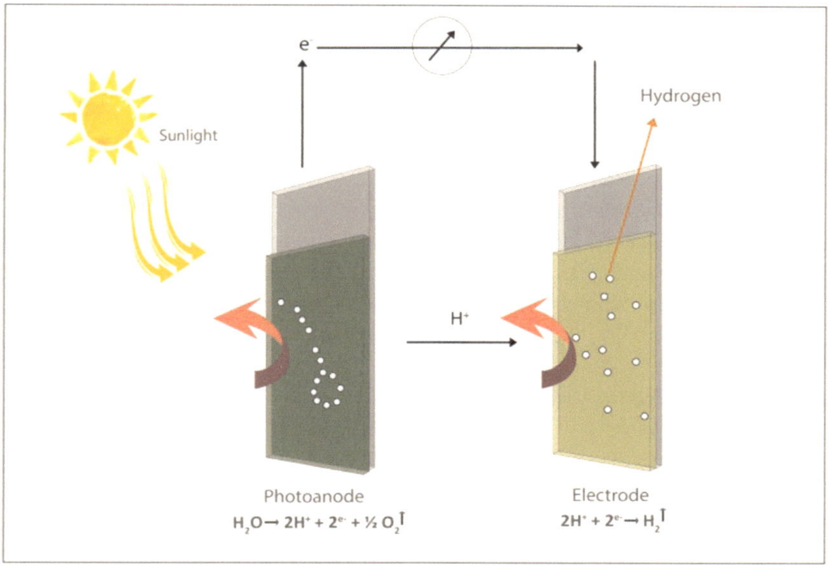

Figure 5.
The working principle of a photoelectrochemical cell where the charge separation is induced by light which leads to water splitting releasing molecular oxygen, proton and electrons. The protons move to the cathode where it is reduced to molecular hydrogen and electrons.

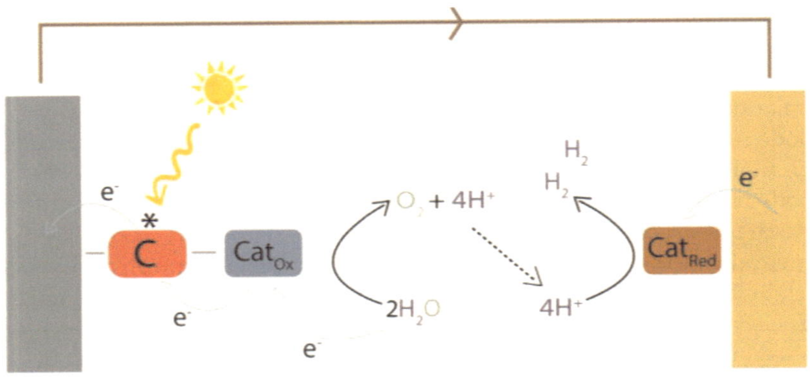

Figure 6.
A modified photoelectrochemical cell having a molecular chromophore and catalyst for water splitting. C is excited chromophore C which transfers electrons and leads to water splitting.*

6.2 Photovoltaic-coupled electrolyzer

A photovoltaic-coupled electrolyzer working principle is based on the working of both solar and electrochemical cells [13, 25]. The photovoltaic cell includes absorption of light as well as charge separation. The energy potential generated by the photovoltaic cell is transferred to an electrolyzer for redox reaction [26]. Thus, in the first step solar radiation is converted into electricity, which is further used for the oxidation and reduction of water or carbon dioxide in the electrolyzer cell. This system has the more efficient strategy and has remarkable solar-to-hydrogen efficiencies of 10–15%, reaching as high as 30%. This ultra-efficient photovoltaic-coupled electrolyzer system utilizes a triple-junction solar cell which is made of three subcells having indium gallium phosphide (InGaP), gallium arsenide (GaAs) and gallium indium nitrogen arsenic antimonideGaInNAs(Sb), respectively. This solar cell involved two polymer electrolyte membranes consisting of a Nafion membrane coated with Pt black catalyst at the cathode and an Ir black catalyst at the anode [27]. The power generated at the triple-junction cell pumped water into the anode of the first electrolyzer and effluent having water and oxygen is further pumped into the second electrolyzer where hydrogen is transferred from the first electrolyzer cathode to the second electrolyzer cathode. As a result, hydrogen and oxygen are collected from the cathode and anode, respectively, at the second electrolyzer, the leftover water is recycled back for the next cycle without any disruption for almost 48 hours with solar to hydrogen conversion efficiency of 30% [27].

7. Semi-artificial photosynthesis

It involves the merging of unique features of both natural and artificial photosynthesis such as high quantum efficiency (100%), selectivity, specificity and self-repair mechanisms of natural photosynthesis [28]. At the same time, the ability to use synthetic material with a wider light absorption spectrum as well as modified molecular chemistries thus minimizing the limitations. The photocatalytic properties of two photosystems PSI and PSII viz., generation of reducing power and photolysis of water, respectively are used for the creation of semi-artificial devices which convert light energy into molecular hydrogen and carbon-based fuels [29].

7.1 Enzyme hybrids

Enzymes have high molecular, chiral and substrate specificity along with high turnover and biodegradability. Enzymes utilization is limited due to their thermal instability, limited optimum pH range, denaturation by organic solvents and inhibition by metal ions. To fully utilize their potentiality as biocatalyst researchers have used porous matrices to elevate their stability *in vivo* (**Figure 7**). Different network materials such as metal-organic frameworks (MOFs) and metal-phenolic networks (MPNs) have been utilized to hold biocatalysts for enhancing their stability. A hundred percent efficiency has been achieved using artificial light absorbers and biological catalysts where photocurrents through electrode-wired enzymes generate kinetically and thermodynamically efficient products at very fast rates. Photosystem II enzymes along with semiconductors and mesoporous opal indium tin oxide (ITO) electrodes are well-studied examples of an inorganic collector with a biological catalyst creating efficient systems [30].

7.2 Cell hybrids

Biological unicellular systems are more efficient in terms of specificity as compared to isolated enzymes from living organisms. The use of microorganisms along with inorganic semiconductors or metal nanoparticles has been investigated by researchers (**Figure 8**) [31]. In one of the studies, the anaerobe *Methanosarcina barkeri*, bacteria, was used along with a materials catalyst, nickel sulfide electrode, a material inspired by naturally occurring nickel-dependent hydrogenases. The bacterial culture was added to the cathode and was exposed to visible light which resulted in the reduction of carbon dioxide to methane with no net loss in performance efficiency [32]. In another study, an acetogenic bacterium *Moorella thermoacetica* was treated with water-soluble gold nanoclusters AuNCs (mainly Au22(SG)18) and illuminated with light of 532 nm wavelength. The cytoplasm mediators in bacteria selectively pick up photo-generated electrons (bypassing cell membrane) and produce acetic acid from carbon dioxide for six days [20]. Another hybrid system was reported where light harvesting indium phosphide (InP) nanoparticles were introduced in genetically engineered yeast cells where photoexcited electrons produced from InP activate nicotinamide adenine dinucleotide phosphate regeneration [33].

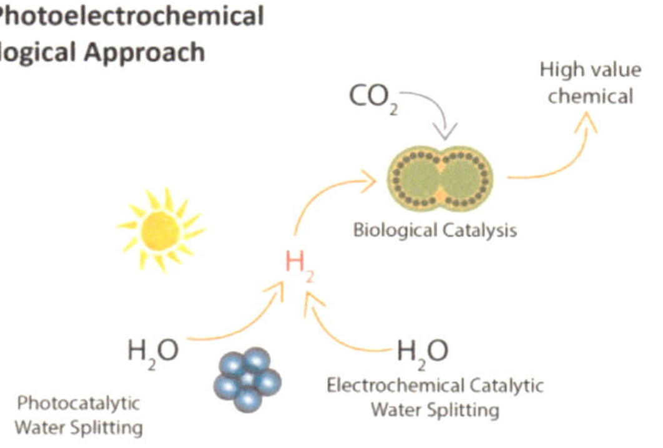

Figure 7.
An outline diagram of a hybrid of photoelectrochemical and biological approach where the biological catalyst is utilized to catalyze the reaction for the production of high-value products.

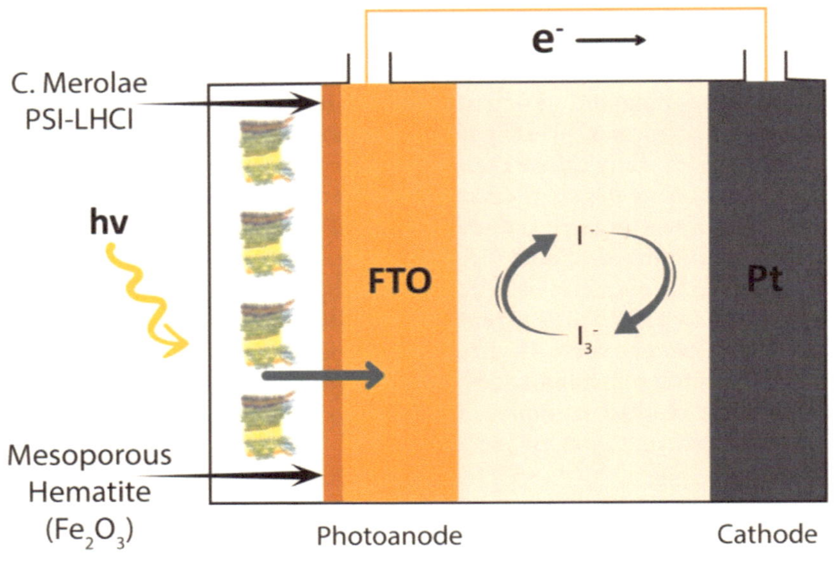

Figure 8.
A biohybrid made by PS-I photosystem I from a red microalga Cyanidioschyzon merolae *nanostructured present on multilayer hematite/FTO electrode.*

7.3 Artificial leaf

The imitation of the Z-scheme and designing of the artificial leaf is another important benchmark in the quest for foolproof operational systems using a biohybrid approach. American chemist Daniel G. Nocera and colleagues (2011) developed a

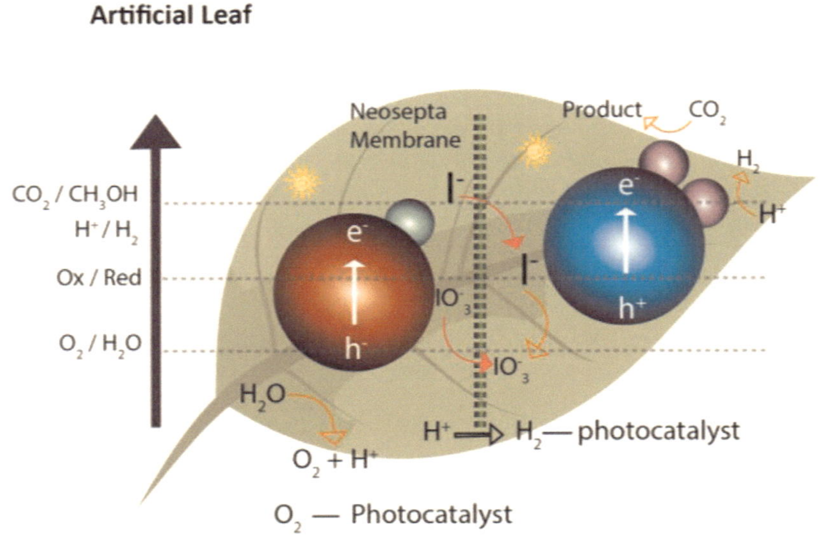

Figure 9.
The artificial leaf designed with a silicon chip coated with a chemical catalyst to speed up the water-splitting reaction on exposure to sunlight. The separated electrons and protons are captured to form hydrogen gas which can be used for the generation of electricity.

silicon-based device that can produce hydrogen and oxygen using solar energy without adding any pollutants was named as artificial leaf [34]. The artificial leaf has a silicon chip coated with a chemical catalyst which on exposure to sunlight speed up the water-splitting reaction and captures separated electrons and protons to form hydrogen gas which can be used for the generation of electricity (**Figure 9**). Artificial leaf still faces significant challenges and focused research is needed to increase hydrogen fuel efficiencies which is very low at about 5% of the total possible fuel availability in solar energy. The technology is highly expensive and the safety of hydrogen fuel generated is negligible limiting the use of artificial leaves for commercial purposes. An artificial leaf was designed using cuprous oxide to produce methanol and oxygen from carbon dioxide and sunlight [35].

8. Improvement strategies

The advancement in nanotechnology (in the area of imaging and modification of nanomaterial) and molecular manipulation such as adding impurities to semiconductors resulted in an increase in light absorption capability, efficient catalyst performance and selectivity [36]. Researchers have created highly efficient antennae complex mimicking light-harvesting complex involved in natural photosynthesis. The antennae reaction center has a hexad nanoparticle having four zinc tetraarylporphyrin molecules, (PZP)3-PZC coupled to a free-base porphyrin-fullerene molecule, P-C60 to form a hexad structured nanoparticle (PZP)3−PZC−P−C60 [37]. The nanoscale materials and devices have been developed by "bottom-up" nanofabrication creating "molecular-lego" which are then used in manufacturing new devices. The creation of supramolecular and their usage in the construction of devices based on molecular components have enhanced the efficiency of catalysts by creating supramolecular cages and hence preventing degradation [17]. The development of single photoelectrode, photovoltaic-coupled electrolyzer, use of nanostructure materials can influence the functioning and efficiency of the device to a greater extent [13, 30]. The physical factors such as temperature, pressure and ion concentration of the environment hold significant leeway.

Wastewater treatment (WWT) leads to the emission of a huge amount of carbon dioxide which is harvested as a source of renewable energy. To tackle the waste-energy-carbon challenge, an integrated approach has been adopted which involves hybrid microbial photoelectrochemical (MPEC) integrated with microbial electrochemical system WWT with artificial photosynthesis. The energy released during WWT aids in achieving carbon neutrality goal side by side assist in solar harvesting, conversion and storage [38].

9. Limitations and challenges

Natural photosynthesis has high quantum efficiency leading to efficient charge separation but the overall conversion of solar energy to chemical energy is quite low nearly about 1%. Using APS, efficiency up to 10% or even higher is demonstrated [12]. One of the most significant bottlenecks in APS is attaining a cost-effective, efficient and stable catalyst material. The organic-based catalyst has the tendency to lose its stability on multiple uses and shows corroding or obstructing the working of system equipment. Many metal-based catalysts have been tried and the search for cost-

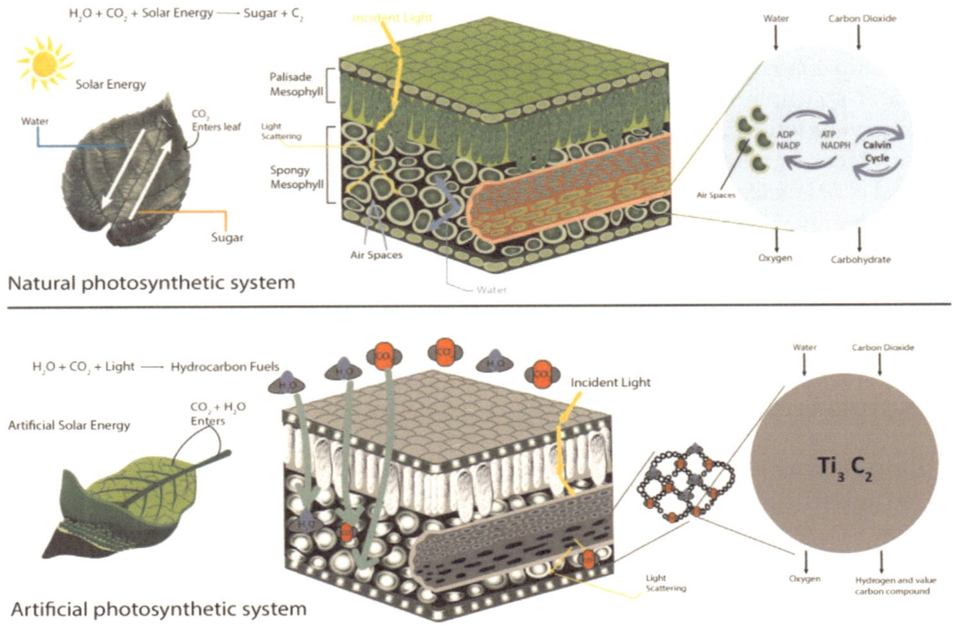

Figure 10.
The comparison of the processes in the natural and artificial photosynthetic system at the leaf, chloroplast and molecular levels.

effective and stability for at least ten years is still under process. To mimic a complex process like photosynthesis is very challenging.

Another notable challenge within the area of mimicking a natural process is the complex molecular geometry found in photosynthesizing organisms. Researchers are having a great deal of trouble replicating the level of intricacy that it entails (**Figure 10**). Many catalysts have been synthesized in the past few decades; however, these catalysts are unstable. Nevertheless, with the help of supramolecular strategies and nanotechnology, scientists are able to easily manipulate the workings of their devices through structural and molecular composition. Studies related to molecular catalyst heterogeneity are limited as it is difficult to match the details present in natural photosynthesis. The development of efficient molecular catalysts will allow the field of APS to advance toward a viable system [15].

10. Future impact and concluding remarks

The top priority of researchers is to search sources of renewable energy that can be used to get some relief from the current state of crisis all over the world [28]. The airplane was created inspired by the flight of the bird, in a similar manner natural photosynthesis serves as a model to mimic the functions of self-sustaining photo-autotroph organisms which hopefully one day create a self-sustaining world. APS already working efficiently and outperforms natural catalytic systems with respect to simplicity, charge transport and light absorption spectral range. Moreover, just as solar panels can be installed onto roofs, providing a secondary source of electricity, future artificial photosynthesis devices can also be applied to power homes as it offers a way to store energy for later use. More than 60% of oil depletion globally is because of its usage in transportation. Electric cars are an excellent alternative but recently new models of cars powered by hydrogen, the byproduct of APS claimed to revolutionize the vehicle industry. These hydrogen-powered vehicles require a very short refueling time and are environmentally friendly.

Natural photosynthesis is solely responsible for all the energy that is required to survive on this planet. In addition, photosynthesis adds energy stored in fossil fuels. It took a billions of years for evolution to make protobiont to evolve into multicellular photosynthesizing systems. To mimic, natural photosynthesis may take more than a decade of extensive research before APS is fully equipped for industrial utilization [11]. Therefore, it is imperative to try and extract energy from a biomimetic approach to this natural process. The endeavor for creating a self-sustained system using APS is still in its infancy. Many successful versions of APS has been devised but not all models are infallible and have drawback related to efficiency, instability or financial expenses. The search for a cost-effective, robust and scalable APS continues in organizations viz., Liquid Sunlight Alliance (LiSA) and Center for Hybrid Approaches in Solar Energy to Liquid Fuels (CHASE). The various attempts to practically apply APS fall short of many efficiencies but still solar fuel production by natural photosynthesis is achievable in the laboratory. The scientific community is well versed in terms and working principles of solar fuel, artificial leaf and artificial photosynthesis and working hard to provide energy using clean, green alternatives globally.

Author details

Renu Kathpalia[1] and Anita Kamra Verma[2*]

1 Department of Botany, Kirori Mal College, University of Delhi, Delhi, India

2 Nano-Biotech Lab, Kirori Mal College, University of Delhi, Delhi, India

*Address all correspondence to: akverma@kmc.du.ac.in

IntechOpen

© 2023 The Author(s). Licensee IntechOpen. This chapter is distributed under the terms of the Creative Commons Attribution License (http://creativecommons.org/licenses/by/3.0), which permits unrestricted use, distribution, and reproduction in any medium, provided the original work is properly cited.

References

[1] Ciamician G. The photochemistry of the future. Science. 1912;**36**(926): 385-394. DOI: 10.1126/ science.36.926.385

[2] Schmidt J, Gruber K, Klingler M, Klöckl C, Camargo LR, Regner P, et al. A new perspective on global renewable energy systems: Why trade in energy carriers matters. Energy & Environmental Science. 2019;**12**: 2022-2029. DOI: 10.1039/C9EE00223E

[3] Tachibana Y, Vayssieres L, Durrant J. Artificial photosynthesis for solar water-splitting. Nature Photonics. 2012;**6**: 511-518. DOI: 10.1038/nphoton.2012.175

[4] Davey T. Artificial photosynthesis: Can we harness the energy of the sun as well as plants? [Internet] 2016. Available from: https://futureoflife.org/2016/09/30/artificial-photosynthesis/ [Accessed: December 15, 2022]

[5] Lal M. Photosynthesis. In: Bhatla SC, Lal M, editors. Plant Physiology, Development and Metabolism. Singapore: Springer; 2018. pp. 159-226. Available from: https://link.springer.com/book/10.1007/978-981-13-2023-1

[6] Fujishima A, Honda K. Electrochemical photolysis of water at a semiconductor electrode. Nature. 1972; **238**:37-38. DOI: 10.1038/238037a0

[7] Maeda K. Photocatalytic water splitting using semiconductor particles: History and recent developments. Journal of Photochemistry and Photobiology. 2011; **12**(4):237-268. DOI: 10.1016/j. jphotochemrev.2011.07.001

[8] Dette C, Pérez-Osorio MA, Kley CS, Punke P, Patrick CE, Jacobson P, et al. TiO_2 anatase with a bandgap in the visible region. Nano Letters. 2014; **14**(11):6533-6538. DOI: 10.1021/ nl503131s

[9] Halmann M. Photoelectrochemical reduction of aqueous carbon dioxide on p-type gallium phosphide in liquid junction solar cells. Nature. 1978;**275**: 115-116. DOI: 10.1038/275115a0

[10] William A. US Patent 4,466,869 Photolytic Production of Hydrogen [Internet]. 1984. Available from: https://patents.google.com/patent/US4466869A/en Accessed: December 11, 2022]

[11] Artificial Photosynthesis. Future Ideas—The Green Age [Internet]. 2017. Available from: https://www.thegreenage.co.uk/tech/artificial-photosynthesis [Accessed: November 15, 2022]

[12] Poudyal RS, Tiwari I, Koirala AR, Masukawa H, Inoue K, Tomo T, et al. Hydrogen production using photobiological methods. In: Subramani V, Basile A, Veziroğlu TN, editors. Compendium of Hydrogen Energy. Volume 1. Hydrogen Production and Purification. Woodhead Publishing Series of Energy, Elsevier; 2015. pp. 289-317. Available from: https://www.sae.org/images/books/toc_pdfs/BELS126.pdf

[13] Yotsuhashi S, Deguchi M, Hashiba H, Zenitani Y, Hinogami R, Yamada Y, et al. Enhanced CO2 reduction capability in an AlGaN/GaN photoelectrode. Applied Physics Letters. 2012;**100**:243904. DOI: 10.1063/ 1.4729298

[14] Layton J. How artificial photosynthesis works [Internet]. How Stuff Works Science. HowStuffWorks. 2020. Available from: https://science.howstuffworks.com/environmental/green-tech/energy-production/artificial-

photosynthesis.htm [Accessed: November 17, 2022]

[15] Balamurugan M, Saravanan N, Ha H, Lee YH, Nam KT. Involvement of high-valent manganese-oxo intermediates in oxidation reactions: Realisation in nature, nano and molecular systems. Nano Convergence. 2018;**5**:18. DOI: 10.1186/s40580-018-0150-5

[16] Paul B, Bhanja P, Sharma S, Yamauchi Y, Alothman ZA, Wang Z, et al. Morphologically controlled cobalt oxide nanoparticles for efficient oxygen evolution reaction. Journal of Colloid and Interface Science. 2021;**582**(A): 322-332. DOI: 10.1016/j.jcis.2020.08.029

[17] Yu F, Poole D, Mathew S, Yan N, Hessels J, Orth N, et al. Control over electrochemical water oxidation catalysis by preorganization of molecular ruthenium catalysts in self-assembled nanospheres. Angewandte Chemie (International Ed. in English). 2018; **57**(35):11247-11251. DOI: 10.1002/anie.201805244

[18] Liu J, Goetjen TA, Wang Q, Knapp JG, Wasson MC, Yang Y, et al. MOF-enabled confinement and related effects for chemical catalyst presentation and utilization. Chemical Society Reviews. 2022;**51**:1045. Available from: https://www.researchgate.net/publication/357731261

[19] Barber J, Tran PD. From natural to artificial photosynthesis. Journal of the Royal Society Interface. 2013;**10**(81):20120984. DOI: 10.1098/rsif.2012.0984

[20] Kumaravel V, Bartlett J, Pillai SC. Photoelectrochemical conversion of carbon dioxide (CO_2) into fuels and value-added products. ACS Energy Letters. 2020;**5**(2):486-519. DOI: 10.1021/acsenergylett.9b02585

[21] Kathpalia R, Verma AK. Bio-inspired nanoparticles for artificial photosynthesis. Materials Today Proceedings. 2021;**45**:3825-3832. DOI: 10.1016/j.matpr.2021.03.214

[22] Styring S. Artificial photosynthesis for solar fuels. Faraday Discussions. 2012;**155**:357-376. Available form: https://pubs.rsc.org/en/content/articlelanding/2012/fd/c1fd00113b

[23] Olmos JDJ, Kargu J. Oxygenic photosynthesis: Translation to solar fuel technologies. Acta Societatis Botanicorum Poloniae. 2014;**83**(4): 423-440. DOI: 10.5586/asbp.2014.037

[24] Braach-Maksvytis V. Nanotechnology and artificial photosynthesis: Go smart, mimic nature. Energy & Environment. 2001;**12**(4): 331-333. DOI: 10.1260/0958305011500814

[25] Zhang B, Sun L. Artificial photosynthesis: Opportunities and challenges of molecular catalysts. Chemical Society Reviews. 2019;**2019**:7. Available from: https://pubs.rsc.org/en/content/articlelanding/2019/CS/C8CS00897C#!divCitation

[26] Gust D, Moore TA, Moore AL. Solar fuels via artificial photosynthesis. Accounts of Chemical Research. 2009; **42**(12):1890-1898. DOI: 10.1021/ar900209b

[27] Jia J, Seitz L, Benck J, Huo Y, Chen Y, Ng JWD, et al. Solar water splitting by photovoltaic-electrolysis with a solar-to-hydrogen efficiency over 30%. Nature Communications. 2016;**7**:13237. DOI: 10.1038/ncomms13237

[28] Xiao K, Liang J, Wang X, Hou T, Ren X, Yin P, et al. Panoramic insights into semi-artificial photosynthesis: Origin, development, and future

perspective. Energy & Environmental Science. 2022;**15**:529-549. Available from: https://pubs.rsc.org/en/journals/journal/ee

[29] Brown KA, King PW. Coupling biology to synthetic nanomaterials for semi-artificial photosynthesis. Photosynthesis Research. 2020;**143**: 193-203. DOI: 10.1007/s11120-019-00670-5

[30] Sokol K, Mersch D, Hartmann V, Zhang J, Nowaczyk MM, Rögner M, et al. rational wiring of photosystem II to hierarchical indium tin oxide electrodes using redox polymers. Energy & Environmental Science. 2016;**9**(12): 3698-3709. DOI: 10.1039/C6EE01363E

[31] Li X, Xu H, Chen Z, Chen G. Biosynthesis of nanoparticles by microorganisms and their applications. Journal of Nanomaterials. 2011;**2011**:1-16. DOI: 10.1155/2011/270974

[32] Paulo LM, Ramiro-Garcia JR, van Mourik S, Stams AJM, Sousa DZ. Effect of Nickeland cobalt on methanogenic enrichment cultures and role of biogenic sulfide in metal toxicity attenuation. Frontiers in Microbiology. 2017;**8**:1341. Available from: https://www.researchgate.net/deref/https%3A%2F%2Fdoi.org%2F10.3389%2Ffmicb.2017.01341

[33] Guo J, Suástegui M, Sakimoto KK, Moody VM, Xiao G, Nocera DG, et al. Light-driven fine chemical production in yeast biohybrids. Science. 2018;**362**: 813-816. DOI: 10.1126/science.aat9777

[34] McAlpin JG, Stich TA, André Ohlin C, Surendranath Y, Nocera DG, Casey WH, et al. Electronic structure description of a [Co(III)3Co(IV)O4] cluster: A model for the paramagnetic intermediate in cobalt-catalyzed water oxidation. Journal of the American Chemical Society. 2011;**133**(39): 15444-15452. DOI: 10.1021/ja202320q

[35] Kim JH, Hansora D, Sharma P, Jang J, Le JS. Toward practical solar hydrogen production – an artificial photosynthetic leaf-to-farm challenge. Chemical Society Reviews. 2019;**48**: 1908-1971. DOI: 10.1039/C8CS00699G

[36] Yang Y, Ajmal S, Zhenga X, Zhang L. Efficient nanomaterials for harvesting clean fuels from electrochemical and photoelectrochemical CO2 reduction. Sustainable Energy & Fuels. 2018;**2**(3): 510-537. Available from: https://pubs.rsc.org/en/content/articlelanding/2018/se/c7se00371d

[37] Kodis G, Liddell PA, de la Garza L, Clausen PC, Lindsey JS, Moore AL, et al. Efficient energy transfer and electron transfer in an artificial photosynthetic antenna-reaction center complex. The Journal of Physical Chemistry A. 2002; **106**:2036-2048. DOI: 10.1021/jp012133s

[38] Li Z, Lu L. Wastewater treatment meets artificial photosynthesis: Solar to green fuel production, water remediation and carbon emission reduction. Frontiers of Environmental Science & Engineering. 2022;**16**:53. DOI: 10.1007/s11783-022-1536-5

Chapter 4

Comprehensive Analyses of the Enhancement of Oxygenesis in Photosynthesis by Bicarbonate and Effects of Diverse Additives: Z-scheme Explanation versus Murburn Model

Kelath Murali Manoj, Nikolai Bazhin, Abhinav Parashar, Afsal Manekkathodi and Yanyou Wu

Abstract

The Z-scheme electron transport chain (ETC) explanation for photosynthesis starts with the serial/sequential transfer of electrons sourced from water molecules bound at Photosystem II via a deterministic array of redox centers (of various stationary/mobile proteins), before "sinking" via the reduction of $NADP^+$ bound at flavin-enzyme reductase. Several research groups' finding that additives (like bicarbonate) enhance the light reaction had divided the research community because it violated the Z-scheme. The untenable aspects of the Z-scheme perception were demonstrated earlier and a murburn bioenergetics (a stochastic/parallel paradigm of ion-radical equilibriums) model was proposed to explain photophosphorylation and Emerson effect. Herein, we further support the murburn model with accurate thermodynamic calculations, which show that the cost of one-electron abstraction from bicarbonate [491 kJ/mol] is lower than water [527 kJ/mol]. Further, copious thioredoxin enables the capture of photoactivated electrons in milieu, which aid in the reduction of nicotinamide nucleotides. The diffusible reactive species (DRS) generated in milieu sponsor phosphorylations and oxygenic reactions. With structural analysis of Photosystems and interacting molecules, we chart out the equations of reactions that explain the loss of labeled O-atom traces in delocalized oxygenesis. Thus, this essay discredits the Z-scheme and explains key outstanding observations in the field.

Keywords: murburn concept, light reaction, photosynthesis, bicarbonate, photolysis, oxygenesis

IntechOpen

1. Introduction to the additive (bicarbonate) sponsored effects

From fossil records, scientists estimate that earth has witnessed about three gigayears of biological photosynthetic activity [1]. Historical documents indicate that by the mid-nineteenth century, the global scientific community was aware that green plants could use sunlight to convert carbon dioxide to carbohydrates, using water and liberating oxygen. By 1930s, Cornelius van Niel had put together a tangible/generic stoichiometry for photosynthesis in diverse organisms utilizing different precursors: $CO_2 + 2H_2A + h\nu \rightarrow CH_2O + 2A + H_2O$, wherein "A" denotes an oxygen or sulfur atom [2]. Although in the early phases of photosynthetic research, it was perceived that O_2 formed originated from CO_2, the use of radioactive isotopes led to the conclusion that water was the source of oxygen [3]. Since the latter half of twentieth century, photosynthesis was traditionally investigated as two separate components: the light and dark reaction. The first part includes NADPH/ATP synthesis and oxygen evolution, whereas the second part comprises the steps that lead to CO_2 fixation. None contend these days that CO_2 serves as the direct substrate in the dark reaction of oxygenic photosynthesis. However, ever since it was reported that bicarbonate ions served to enhance oxygen evolution in the light reaction [4], the photosynthetic research community has been divided on the role(s) of this seemingly innocuous species. While some believe that the stimulatory outcome is purely due to indirect effects on Photosystem II functioning (where bicarbonate has been reported to bind at different loci) [5, 6], some have advocated a more direct, that is, substrate role of bicarbonate [7, 8]. In later times, as some researchers emphatically claimed that there was no evidence for the direct roles of bicarbonate [9–12], the general opinion shifted against the "PS II-binding based effects" school of thought. In response, while a few of the original advocates toned down their support for bicarbonate being crucial [13], a few others have campaigned steadfastly for bicarbonate to play more direct roles [14–16]. In the context of the conundrum prevailing on bicarbonate (as well as several other additive molecules'/ions' ability to impact the photosynthetic process), we apply the murburn model of photosynthesis [17–25] to clarify upon several other associated facts from the literature and reason out the structure–function correlations of proteins and dynamics of molecular interactions in milieu.

2. Theoretical approach to a convoluted problem

In this theoretical exploration/analysis, we survey the potential reactions occurring in milieu with a probabilistic approach. We theorize that if affinity-based interactions drive electron transfers in physiology, the interaction dynamics between donor-acceptor pair must be governed by the various known molecular descriptors of affinity. Further, in continuance with our earlier works in the field of bioenergetics, we aim to understand and predict the feasibility of electron/group transfer processes based on the reactant's/reaction's thermodynamic profiles and known aspects of kinetics. Details on the various modalities of thermodynamic calculations are reported elsewhere [26, 27]. These insights are then applied to study the structure–function correlations of Photosystem II and other biomolecules of the photosynthetic reaction system in thylakoids. The known structure of cyanobacterial Photosystem II (pdb:6JLJ) [28] (and other proteins/biomolecules) was analyzed by visualization software like Chimera 1.12 [29] and PyMOL [30]. Cavities of proteins were analyzed by

POCASA [31]. Small molecular/ionic properties were availed from ChemSpider/PubChem registry and ACDLabs/ChemAxon predictions.

3. Addressing experimental observations with the classical Z-scheme and murburn models

3.1 Evident flaws in photosynthetic "electron flow" and their explanations

From the earlier perceptions that photo-assisted reactions directly resulted in complex group transfers, the consensus had set in by the mid-later part of twentieth century that light reaction led to a deterministic series of transfers of electrons derived from water bound at Mn-complex of PS II (along intermittent pit-stops at the redox centers of multiple proteins/biomolecules), leading ultimately to the reduction of $NADP^+$ bound at flavoenzyme reductase. This gradual outer-sphere electron transfer [32] process was supposedly synchronous with "pumping of protons," whose energy was subsequently harnessed for making ATP. The ability to fix CO_2 in the dark reaction rested on the redox/phosphorylation power of NADPH/ATP formed in the light reaction.

One crucial oversight made in the earlier times was that the photosynthetic product of molecular oxygen (an omnipresent molecule in thylakoids) could also serve as an electron acceptor in the Hill (or light) reaction [33]! The diradical of oxygen is both 1e and 2e active and it is difficult to imagine any deterministic regulatory mechanism that could override shunting by oxygen. Further, it was/is common knowledge that several ions and natural/synthetic redox-active molecules could affect or donate/accept electrons with respect to the photosynthetic ETC [34–41]. In the context of the current discussion, this important aspect is investigated further. A compilation (from the publications referenced above and the citations mentioned therein) of the affinity descriptors of some of the spectrum of ions/molecules known to shunt/affect the photosynthetic pathway by serving as electron *donors or donors and acceptors (both)* to

No.	Name (projection radius)	Molecular Formula	Mass (g/mol)	To PSI/PSII	MV (cm^3)	PSA ($Å^2$)	Log P	H-Bonds	Rotating bonds
1	Tetraphenylboron (tetraphenylborate)	$C_{24}H_{20}B$	319	I/II	na	na	na	0.0	4
2	Diphenylcarbazide	$C_{13}H_{14}N_4O$	242	II	187	65	2.24	5.4	4
3	Ferrocyanide	C_6FeN_6	212	II	na	143	na	6.0	0
4	2-ketogluconate	$C_6H_9O_7$	193	II	na	138	−2.82	7.5	5
5	Benzidine	$C_{12}H_{12}N_2$	184	II	159	52	1.5	2.4	1
6	Hydrazobenzene	$C_{12}H_{12}N_2$	184	II	156	24	2.94	2.2	3
7	Ascorbate (3.9–5.4)	$C_6H_7O_6$	175	I/II		110	−2.41 (−1.26)	6.4	2
8	TMPD, Wurster's blue (3.5–5.9)	$C_{10}H_{16}N_2$	164	II	165	6 (6.5)	2.08	2.0	2
9	Iodide	I^-	127	II	na	na	na	na	na
10	Cysteine (3.0–4.5)	$C_3H_7NO_2S$	121	II	91	102 (63)	0.23 (−2.79)	3.3	2

No.	Name (projection radius)	Molecular Formula	Mass (g/mol)	To PSI/PSII	MV (cm^3)	PSA ($Å^2$)	Log P	H-Bonds	Rotating bonds
11	Hydroquinone (3.4–4.3)	$C_6H_6O_2$	110	I/II	86	40	0.64 (1.37)	2.2	0
12	Phenylenediamine (3.4–4.6)	$C_6H_8N_2$	108	I/II	94	52	0.05 (0.32)	2.4	0
14	Semicarbazide (2.8–3.7)	CH_5N_3O	75	II	43	82	−1.61	4.5	1
15	Hydrogen peroxide (2.0–2.3)	H_2O_2	34	II	24	40	−0.43	2.2	1
16	Hydroxylamine (2.0–2.6)	H_3NO	33	I/II	30	46	−0.81	2.3	1
17	Hydrazine (2.1–2.6)	H_4N_2	32	I/II	36	52	−1.2	2.4	1
18	Water (1.7–2.0)	H_2O	18	II	18	(25.3)	−1.38 (−0.65)	1.2	0
19	*Bicarbonate (2.6–3.0)*	*CHO_3*	*61*	*??*	*na*	*60*	*−0.81 (0.25)*	*3.2*	*0*

For smaller molecules, values in braces are from ChemAxon (given when significant differences exist with ACDLabs). H-bond entries are the number of acceptor atoms and donor atoms, respectively. MV and PSA stand for molar volume and polar surface area, respectively.

Table 1.
Affinity descriptors of various known electron donors to Photosystem I and/or Photosystem II.

No.	Name	Molecular Formula	Mass (g/mol)	To/From PSI/PS II	MV (cm^3)	PSA ($Å^2$)	Log P	H-Bonds	Rotating bonds
1	TMPD (3.5–5.9)	$C_{10}H_{16}N_2$	164	II	165	6	2.08	2.0	2
2	Menadione	$C_{11}H_8O_2$	172	I	141	34	2.38	2.0	0
3	Phenazine	$C_{12}H_8N_2$	180	I	144	26	2.84	2.0	0
4	Diquat	$C_{12}H_{12}N_2$	184	I	na	8	−4.71	2.0	0
5	Pyocyanine	$C_{13}H_{10}N_2O$	210	I	na	40	na	3.1	1
6	Diaminobenzidine	$C_{12}H_{14}N_4$	214	I/II	164	104	−0.95	4.8	1
7	Paraquat (methyl viologen)	$C_{12}H_{14}Cl_2N_2$	257	I	na	7.8	1.7	0.0	1
8	Dichlorophenol indophenol (DCPIP)	$C_{12}H_7NCl_2O_2$	268	I	187	50	1.91	3.1	1
9	Flavin mononucleotide (FMN)	$C_{17}H_{22}N_4NaO_{10}P$	496	I/II	na	214	−1.7	10.3	6
10	Plastoquinone (PQ)	$C_{53}H_{80}O_2$	749	I/II	808	34	20.18	2.0	26

Table 2.
Some affinity descriptors of various known electron donors and acceptors (both!) to PS I and/or II.

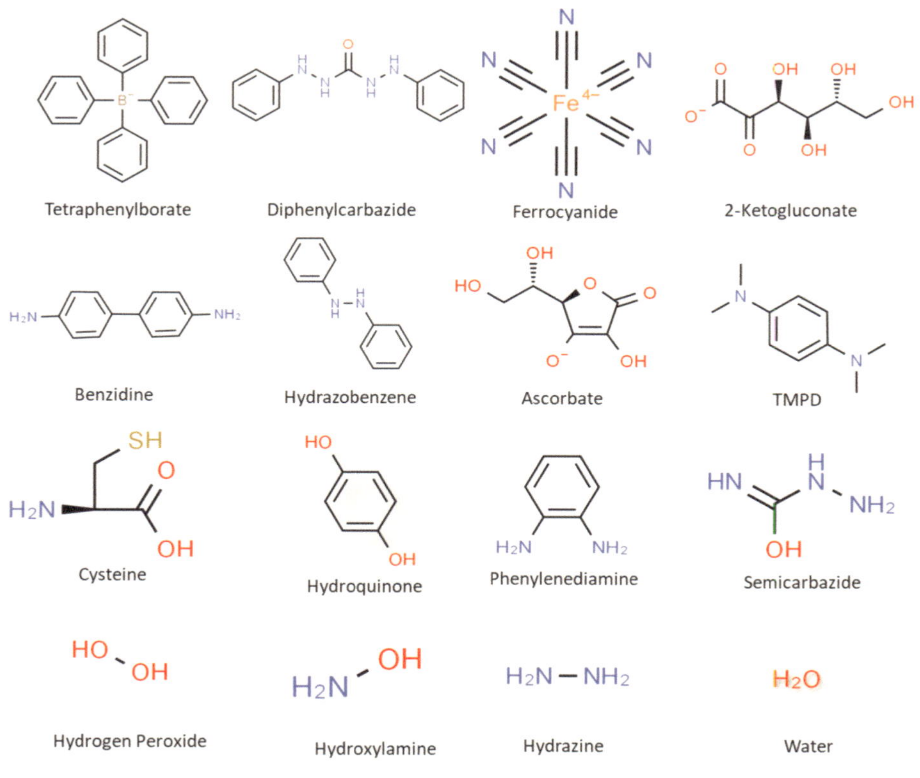

Figure 1.
Structures of molecules and ions serving as electron donors to PS I/II.

the two photosystems is presented in **Tables 1** and **2** and **Figures 1** and **2**, respectively. (In continuum, **Table A1** and **Figure A1** of Item 1 of Appendix lists a compilation of some electron acceptors from PS I and II).

It can be noted and inferred that if such a diverse group of chemical species could serve as source of electrons to PS I/II, it is quite probable that bicarbonate could also serve in this role. Clearly, there is little selectivity/specificity in the overall e-transfer mechanism, as these molecules differ in drastic ways. Also, it is difficult to accept that such diverse molecules posing discrete descriptors of affinity/reactivity (varying topographies, geometries/projection radius/surface area/volume, electrostatic signatures, redox potentials, partition coefficients, hydrogen bonds, rotatable bonds, etc.) could proffer impacting outcomes, if the physiological electron transfer mechanism were to be based on affinity-based interactions between the so-called donor-acceptor pairs, at defined loci. Some molecules are seen to donate and accept electrons to the same photosystem or to both photosystems; this would surely not afford any directionality. Therefore, the findings clearly suggest that deterministic ETC (such as Z-scheme) must be discounted; as any such serial electronic circuitry would not work sustainably in physiology. For this to happen in some miraculous ways, oxygen and several other reaction components must somehow not behave in their "natural" way! In this regard, we have pointed out that the same crucial oversight was made in

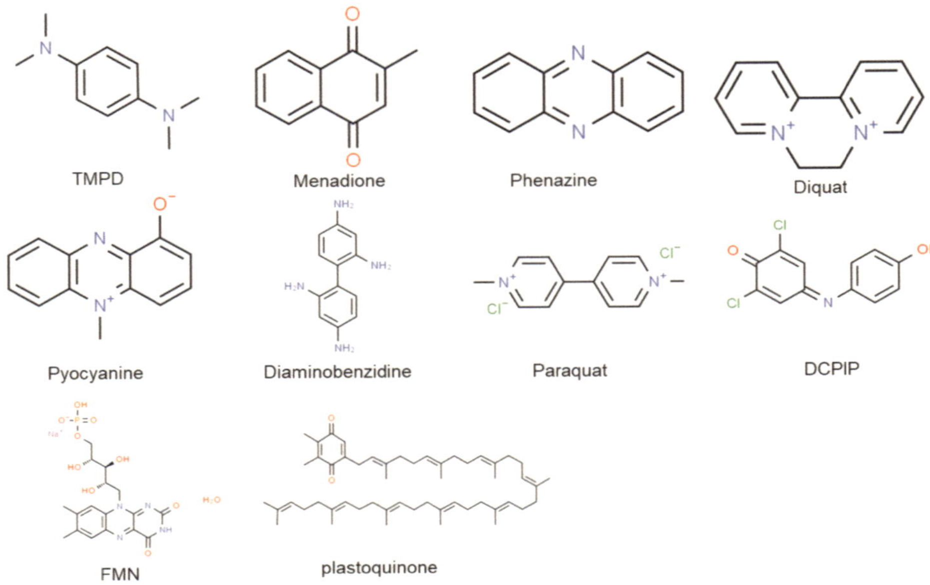

Figure 2.
Structures of molecules and ions serving as electron donors and acceptors (both!) to PS I/II.

respiratory physiology. In the mitochondrial and molecular respiratory system of various cells, several molecules and ions of diverse geometries, dimensions, and redox potentials could also serve as electron donors/acceptors [26]. We had reasoned this fact with the spontaneous murburn equilibriums occurring in milieu.

Further, the classical ETC-CRAS explanation for photosynthesis would also mandate the following insurmountable premises:

i. Several dozens of deterministic e-transfer steps must occur before the release of an oxygen molecule at the WSC. {For example, a dozen e-transfer steps must occur in a single Q-cycle of four electrons- $[(QH_2 - FeS - Heme - PC = 3) \text{ x } 2] + [(QH_2 - Heme - Heme - Q = 3) \text{ x } 2] = 12$. As only four electrons reach PC for every 8 electrons going through Q cycle, the Q-cycle component alone would take two dozen e-transfer steps for the release of an oxygen molecule!}. It is impossible to imagine the orchestration of such a fastidious outcome, given that several components are distributed at unfavorable ratios and not arranged in the sequence or locations where mandated.

ii. Non-existent/unavailable protons are needed to build the proton motive force for serving the endergonic ATP synthesis {For example, a cyanobacterium of 0.5-micron dimension has a volume of $\sim 0.125 \times 10^{-15}$ liters. The usual functioning of these cells occurs at a pH of 8, wherein protons are at 10^{-8} M concentration. Since a liter of pH 8 solution has $6.023 \times 10^{23} \times 10^{-8}$ protons (= $\sim 6 \times 10^{15}$ protons), the small volume of a cyanobacterium has only $0.125 \times 10^{-15} \times 6 \times 10^{15}$ = 0.75 protons! This is when there are tens of thousands of protein complexes (Complexes I–V) in a cell or bioenergetic organelle.

iii. Even if protons are made available through some miraculous means, how is the directionality of ATP synthesis by complex V ascertained? As per the Boyer model, proton moving in or out via the c-ring determines the hydrolysis or synthesis at the alpha-beta subunits. Since there is little proton gradient in physiology, how does a proton-based rationale give ATP synthesis when protons are not used at the active site and when protons are present in both in- and out- phases [20]?

iv. Perhaps, the arrangements of components/pigments in the photosynthetic structures in earlier revealed bacterial systems could have indicated an ordered mechanism [42]. However, the structural distribution of plant pigments, photosystems, and LHC arrangement in chloroplasts revealed later show very little order [43]. Therefore, energy transfer between the various light-absorbing photo-active pigments scattered around in the thylakoid membranes and the redox-centers of photosynthetic proteins must occur via stochastic measures [23].

v. The equations prescribed for the overall process must violate the fundamental laws of thermodynamics for attaining viability. {$2NADP^+ + 3ADPOH + 3POH \rightarrow O_2 + 2NADPH + 2H^+ + 3ADPOP + H_2O$; Overall $\Delta_r G\prime^o{}_{aq} = 1463.8$ kJ/mol) is the prescribed equation. The input for the overall process is 4 einsteins of 680 nm photons ($4\times175.9 = 703.6$ kJ/mol) and 4 einsteins of 700 nm photons (4×170.9 kJ/mol = 683.6 kJ/mol), giving a total of only 1387.2 kJ/mol, which means that there is a significant shortage of 76.6 kJ/mol to even things out energetically!}

Daniel Arnon, the pioneer who laid the foundation of cyclic/acyclic photosynthesis and made key contributions that led to the establishment of Z-scheme [44–46] changed his perception and tried to "wade against the current" by quoting several arguments and demonstrating NADPH formation in ways other than the Z-scheme format ([47–49] and several citations from his group mentioned therein, starting from 1980). Z-scheme's steadfast adherents chose to sidetrack Arnon and the undeniable evidence which showed that crucial components (such as plastocyanin and cytochrome b_6f) of the deterministic e-transfer scheme were in fact, completely optional in physiology [50–52]. Therefore, such conclusive evidence dictates that the classical ETC-CRAS model for light reaction of photosynthesis should be jettisoned, and it is in this context that murburn model presents a viable alternative [19, 22].

3.2 Asking the right question: How do the photosystems work?

Since the electron flow is deterministic in Z-scheme, Photosystem II (PS II) and Photosystem I (PS I) must have a primary donor and acceptor site each, called D2-A2 and D1-A1, respectively. That is, Z-scheme dictates that the small molecule of water donates electron at D2 and the biomolecule of quinone accepts at A2, whereas the protein plastocyanin donates at D1 and the protein ferredoxin accepts at A1 (**Figures 3** and **4**). In this regard, the information that photosystems can give/accept electrons to/from a wide variety of redox-active small molecules and ions (as shown in the earlier section) is very crucial. From a survey carried out in our study of available literature and seconded by the earlier insightful assertions of Hauska [35], a statistical criterion is

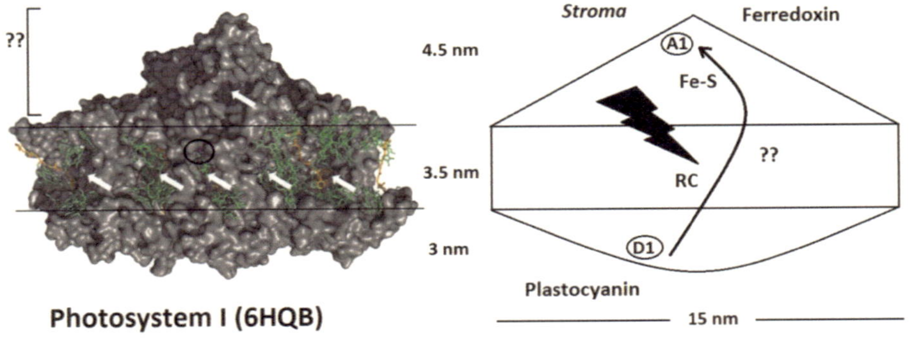

Figure 3.
A schematic representation of the Z-scheme functionality of Photosystem I, drawn to approximate scale. The left panel shows the structure of PS I and the right panel shows the essential schematic representation of the same. As seen, there are several chlorophylls and carotenoids scattered in the membrane-phase of PS I (marked out with the horizontal straight lines) and the stromal phase apoprotein has Fe-S centers. The white arrows mark grooves/ cavities in the protein, enabling diffusible species dynamics. It is not clear how the small amount of plastocyanin could provide electrons to the electron-deficient RC, post photo-activation induced e-transfer processes. Also, some of the FeS centers and the overall structure of stromal part of PS I are not functionally accounted by Z-scheme perceptions [19, 22]. The question mark in the left and right panels pose queries on the roles of the large extra-membrane apoprotein and the mechanism of deterministic electron relays, respectively. The circled spot in the left panel is the reaction center (RC).

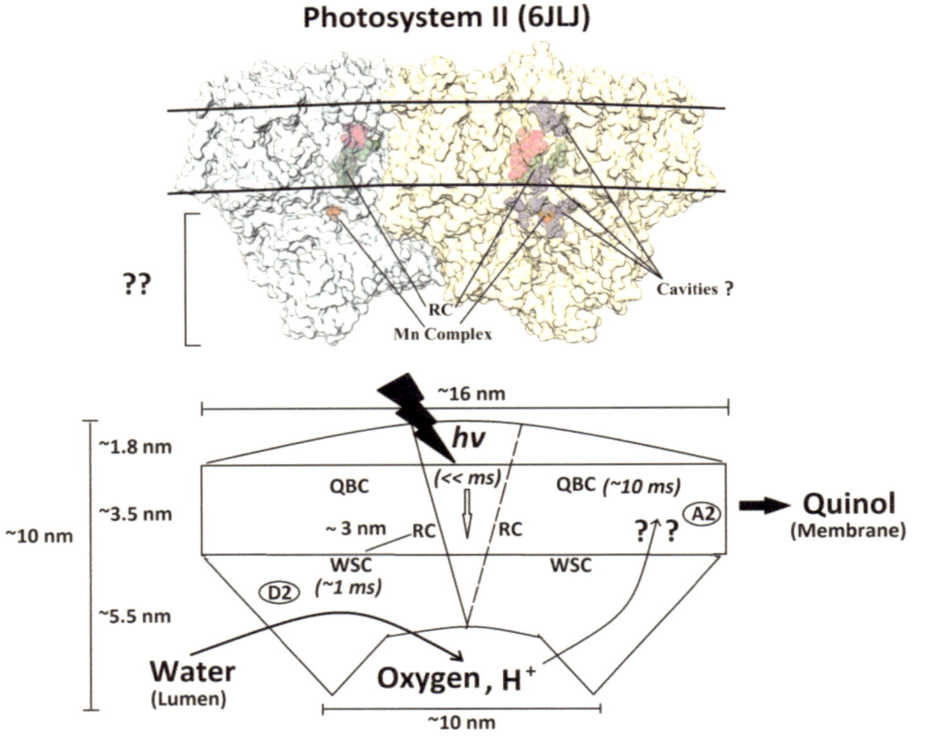

Figure 4.
A schematic representation of Z-scheme functionality of PS II, drawn to approximate 2D-scale. (Please refer text for discussion.) The purported binding site of bicarbonate is the non-heme iron center (located near the pointed arrow of the lightening sign in the bottom panel image. The big question marks on the left of the top panel ask- why should PS II have such bulbous extensions? The cavities in the PS II structures are inexplicable in the classical purview.

evident. Hydrophilic ions or molecules (such as ferricyanide or benzoquinone-2-sulfonate) are served at A1, whereas hydrophobic molecules (such as benzoquinone or p-phenylenediamine) are served at A2. Although the diversity of acceptors is not agreeable to the affinity-based binding rationale (deemed necessary for selective electron transfers), the location of the acceptor sites apparently seems to be in alignment with the Z-scheme. It is crucial to note that while lipophilic quinones could serve at D1 (and D2), water-soluble compounds (like sulfonate derivatives of phenazine or DCPIP) could not. This finding directly goes against the Z-scheme layout, because in both these photosystems, electrons are supposed to be availed from the soluble phase (plastocyanin in PS I and water in PS II). It can be seen from the earlier section that synthetic/natural molecules like TMPD/plastoquinone are known to give and accept electrons to/from both photosystems! How is it determined in physiological premises whether a molecule could serve as a donor or acceptor, and at which port? Particularly, while some researchers disown that bicarbonate has any binding-based physiological role in PS II function at one hand [12] (which others quote to argue that the stimulation by bicarbonate is an artifact!), yet others claim that bicarbonate is an essential binding-based cofactor for PS II [53]. How can the feud be settled? The answer is unavailable in the classical perspective, which perceives only binding-based effects as important criteria. Therefore, the right question to be asked is- what is the role of the photosystems in the light reaction? Or, how do the diverse molecules impact the photosystems' functioning?

3.3 Murburn model for the light reaction of photosynthesis

Today, we know that Z-scheme is unsuitable for explaining the synergy Emerson observed between Photosystems I and II [54–56]. The fatal logical flaw is that a serial arrangement of components would surely lower e-flow/reaction rates (thereby negating the logic of Z-scheme conception!) and only a parallel functioning of components could explain the synergy of photosystems [19]. Therefore, the imperative mandate for venturing beyond the classical paradigm of "Kok-Joliot cycle (KJC), Z-scheme ETC, Q-cycle and chemiosmotic rotary ATP synthesis (CRAS)" must be registered [18]. In this regard, we have conclusively critiqued the classical view [18, 57], rendering it incapable of redemption. In lieu, we have proposed a murburn concept based explanation for oxygenic photosynthesis (**Figure 5**), which is based on DR(O)S mediated catalytic outcomes.

The new model (which need not depend crucially on binding-based outcomes but is more an interactive dynamics of molecules, unbound ions, and reactive radicals in milieu) justifies the structure of proteins, architecture/distribution of components and organelles, overall thermodynamics, kinetics, probability considerations and affords a globally valid/tangible mechanistic explanation for the light reaction [17, 19, 21–26, 58]. Under this scheme, the photosystems enable ECS (effective charge separation) and facile charge replenishment (**Figure 6**). Further, electron transfers in this scheme need not be based solely on donor-acceptor affinity binding-based interactions (**Figure 7**). It is in this context of murburn model that we address the long-standing debate on the role (s) of bicarbonate ions (and other non-specific agents) on the dynamics and efficacy of the light reaction. From the two **Figures 6** and 7, it can be seen that the stochastic murburn model permits electron inputs/withdrawals via the involvement of unbound ions and oxygen. This is enabled by the generation of transient electron-charged or electron-deficient species from these agents, which generate a pool of redox relays. The

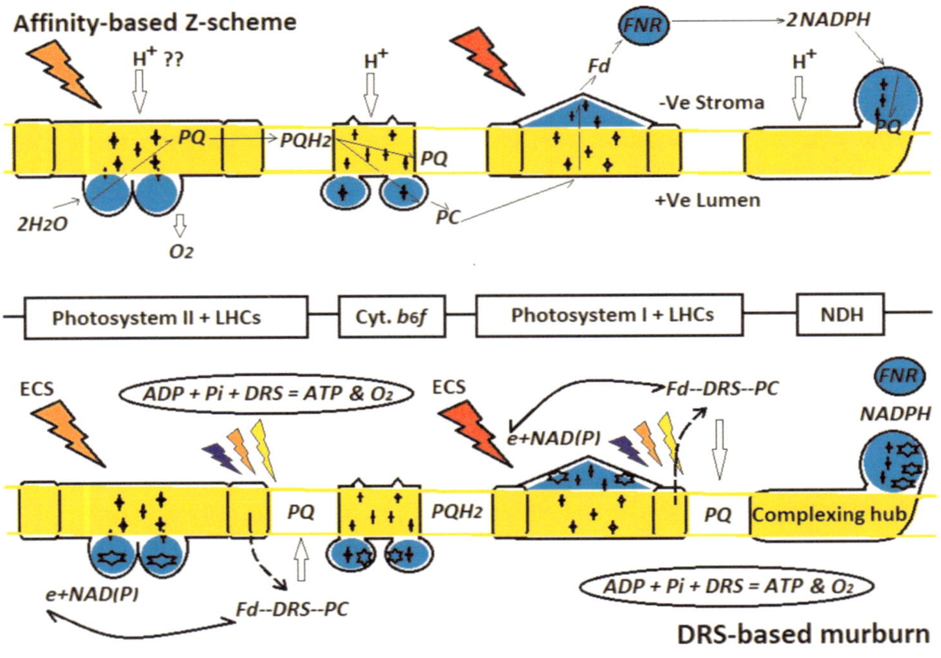

Figure 5.
Structure–function correlations of chloroplast/thylakoid membrane-embedded redox proteins, under Z-scheme and murburn models. The redox centers (porphyrins) in protein complexes are designated with '+' sign. In the Z-scheme shown on top, LHCs are exciton relay agents, conveying photons of two specific wavelengths to the two Photosystems II & I, which generate the mobile reducing equivalents of plastoquinol (PQH2) and reduced ferredoxin (Fd) from water and reduced plastocyanin (PC) respectively. This is when Cytochrome b_6f and NADPH-dehydrogenase (NDH) serve the role of proton pumps, to generate a proton motive force or trans-membrane potential (TMP), which is harnessed by F_oF_1ATPase to make ATP by the CRAS mechanism. Only PS II utilizes water and produces oxygen in this scheme and NADPH is produced by FNR at the end of a multi-dozen step e-transfer process. All components are mandated to have a definite order of electron donor and acceptor function, arranged in a series without any role for oxygen or DRS. Stoichiometry is depicted only for the formation of one molecule of oxygen from two molecules of water, giving rise to two molecules of NADPH. The stroichiometic requirement/involvemnt of other intermediates and formation of ATP are not shown. In the murburn model shown on the bottom, all photo-active pigments are involved in redox reactions, whereas only PS II & I can bring about effective charge separation (ECS). Due to photo-activated liberation of electrons in the form of diffusible reactive species (DRS), soluble proteins Fd and PC are involved in redox equilibriums with DRS. Similarly, the binding sites for ADP (shown has six-cornered stars) on the membrane protein complexes enable effective activation of ADP/Pi the photophosphorylation process. Nicotinamide reduction is via simpler bimolecular processes requiring the buffering of Fd. Oxygen involvement and evolution in the murburn model is delocalized, although PS II's Mn-complex can serve as an effective peroxidase, driving higher oxygen evolution. Since the murburn model is inherently a stochastic/statistical model (as a result of an aggregate of several parallel competing reactions), stoichiometry is variable and non-integral. For the facts/arguments which establish conclusively that Z-scheme CRAS is untenable (besides the ones presented in the paragraph leading to the introduction of this figure) and for a greater clarity on the murburn model for light reaction, please refer our recent publications [19, 22].

formation of stable 2e products and their utilization and/or porting thereafter determines the reaction dynamics in milieu.

3.4 Thermodynamics and kinetics of physiological bicarbonate reactions

In plant and animal cells, the hydration/dissociation equilibrium of CO_2-bicarbonate system is of immense physiological significance and therefore, well-studied. When gaseous CO_2 mixes with water, there are four particles/species formed that co-exist in a complex interactive equilibrium: CO_{2aq} (a, dissolved/aquated molecule), H_2CO_3

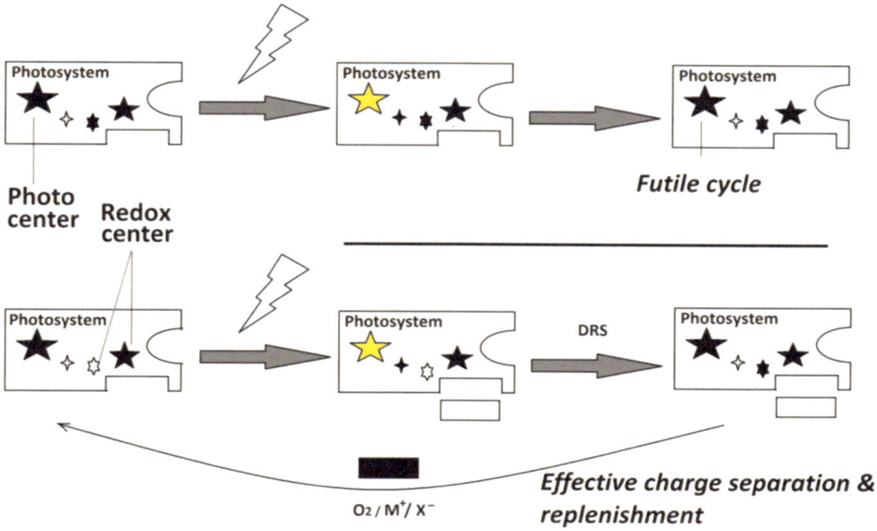

Figure 6.
A schematic to explain the effective charge separation principle of murburn concept. A futile cycle is shown in the top panel wherein photoactivation (thunderbolt sign) of a center releases an electron which is retrieved at the same center, without any other event occurring in the milieu. In the lower panel, the arrangement of suitable redox centers in proper redox states and the involvement of DRS ensure effective charge separation, wherein the photoelectric electron does not go back to its original source. The ECS at a photosystem enables NADP reduction and photophosphorylation thereafter. Here, DRS are shown to replenish the electron-deficient photo-active center. This outcome could also occur through tunneling, if a suitable redox center is available nearby. M and X stand for cation and anion species, respectively. Blackened stars or rectangles represent reduced (e-replenished) states whereas yellow stars represent photoactivated center that has lost an electron.

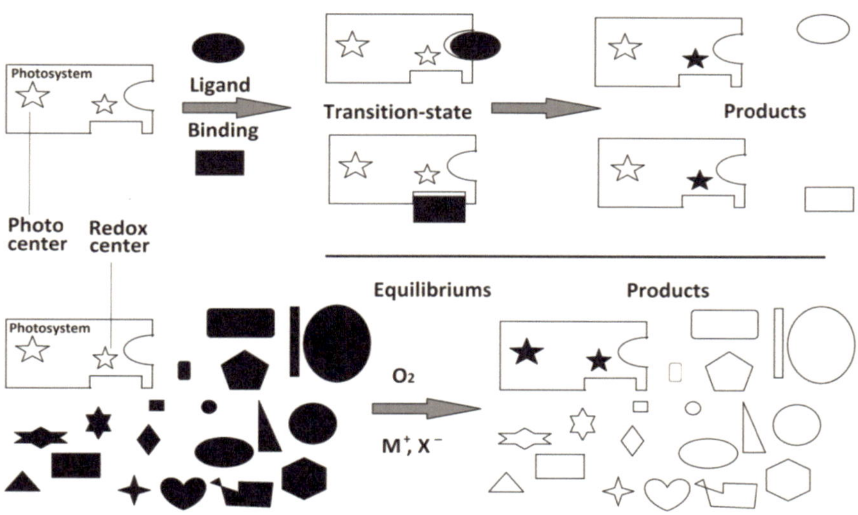

Figure 7.
The murburn modes of electron transfer: While the classical scheme (top panel) only entails affinity driven ET between donor-acceptor complementation (allowing receipt of electrons at the redox center close to the binding site), the murburn model does not negate this possibility but also endorses interactive equilibriums involving diffusible reactive species (DRS). It can be noted that the involvement of DRS enables the electron transfers between remote redox centers with photo-active centers and molecules/ions of diverse topography and other features. Blackened shapes represent reduced states whereas unfilled shapes represent oxidized states.

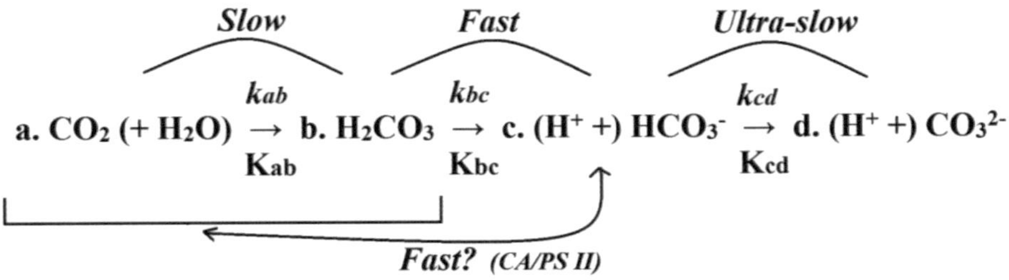

$$K_{ab} = k_{ab}/k_{ba}\ ;\ K_{bc} = k_{bc}/k_{cb}\ ;\ K_{cd} = k_{cd}/k_{dc}\ ;\ K_{abc} = K_{ab}/(1 + K_{bc})$$

Figure 8.
Simplified representation of the highly complex "carbon dioxide-carbonic acid-bicarbonate-carbonate" interactive equilibriums and the relevant kinetic/thermodynamic constants. The direct a-c equilibrium is too slow and c-d equilibrium is relevant only at very high pH. (The constants involved for a-b are secondary derivations. Only the forward rate constants are depicted with italicized "k" on the upper side of the directional arrow, whereas the equilibrium constant is depicted with capitalized "K" on the lower side.) The relations between kinetic and thermodynamic constants are given below the equations. (https://www.aqion.de/site/carbonic-acid-kinetics#fnref:1).

(b, uncharged carbonic acid), HCO_3^- (c, monovalent bicarbonate anion) and CO_3^{2-} (d, divalent carbonate anion). The scheme of these species and the equilibrium/kinetic constants governing their interactions is presented in **Figure 8**. Although the enzyme carbonic anhydrase (CA, a zinc cofactor containing enzyme present at high copy number in chloroplasts) has been extensively studied and reviewed periodically [59–61], it is not really clear as to how these interactive equilibriums operate in physiological dynamics. Also, it has been a general understanding that Photosystem II (PS II) has some CA-type activity [62, 63]. The theoretical investigation bears relevance in light of renewed interest in the roles of bicarbonate and CA [64]. In this context, we would like to point out that perceptions involving the roles of protons in overall thermodynamic treatments were misplaced earlier [27]. Therefore, a revisit to the pertinent treatment is mandated in the new light of awareness.

3.4.1 Well-studied 2e reactions and their equilibriums

The overall interactive process and equilibriums can be understood by considering formative steps for the three soluble ingredients: carbonic acid, bicarbonate ion, and carbonate ion. We consider the theoretical and experimental information on individual reactions from the complex equilibriums. Please consult Item 2, Appendix, for the details of calculation of the empirical $\Delta_r G°$ (standard free energy change of reaction) from the $\Delta_f G°$ (standard free energy change of formation) values, which are given within braces in the pertinent equations. The values determined in this study are given in the first line of the equation and those seen/sourced from literature (as listed in Item 2, Appendix) are in the next line. The pitfalls of experimentally determined equilibrium constants are discussed elsewhere [27].

3.4.1.1 Formation of carbonic acid

$$CO_2\ (-386) + H_2O\ (-237.2) \rightarrow H_2CO_3\ (-623.1);\ \Delta_r G° = 0.1\text{kJ/mol};\ \approx K_{eq} = 0.96\text{M}^{-1}$$

$$\text{Exp.}K_{eq} = [H_2CO_3]/[CO_2] \times\ [H_2O] = 1.3 \text{ x } 10^{-3} \text{ or } 5.4 \text{ x } 10^{-5}\ (\approx 16.5\text{kJ/mol or } 24.3\text{kJ/mol})$$

The rate constants were 0.039 s^{-1} for the forward reaction and 23 s^{-1} for the reverse reaction [65], which gives K_{eq} to be 1.7×10^{-3}. In another study, the rate constants were 0.04 (or 0.025 to 0.04) s^{-1} for the forward and 18 (or 10 to 20) s^{-1} for the backward reaction [66]. The second order forward rate constant is 0.0027 $M^{-1}\,s^{-1}$ (which multiplied by 55.5, the molarity of water, gives the first order reaction rate constant as 0.15 s^{-1}) whereas the reverse reaction is 50 s^{-1}. Thereby the equilibrium constant is 0.0027/50 = $5.4 \times 10^{-5}\,M^{-1}$ [67] or the dimensionless constant would be 3×10^{-3} (= 0.15/50). Therefore, the major amount of carbon-dioxide remains as soluble gas and does not get hydrated to become an acid. In physiological ranges of pH and temperature, the ratio is ~340 CO_2: 1 H_2CO_3. Though there is a considerable spread in the literature regarding the experimental kinetics and equilibrium values, the consensus understanding is that this equilibration process is relatively slow, and the hydration/dissociation process can go both ways, with hydration being less preferred. Very importantly, although the overall directional trend predicated by empirical/theoretical energetics and experimental considerations are in agreement, the value of $\Delta_r G°$ does not correspond accurately to experimental K_{eq} (as expected from the equation: $\Delta G = -RT\, ln.K_{eq}$) in this CA mediated primary reaction (which is theoretically independent of pH, as evident in the equation).

3.4.1.2 Formation of bicarbonate ion

As per **Figure 8**, we can envisage three interconnected ways, 2p, 2q, and 2r.

$$CO_2\ (-386) + H_2O\ (-237.2) \rightarrow H^+\ (412.5) + HCO_3^-\ (-999.8);$$
$$\Delta_r G° = 35.9\ \text{kJ/mol};\ \approx K_{eq} = 5.1 \text{ x } 10^{-7}\ (2p)$$

$$\text{Exp.}K_{eq} = [H^+] \times\ [H_2CO_3]/[CO_2] \times\ [H_2O] = 1.5 \text{ x } 10^{-8};\ (\log\ K = -7.82)$$

$$\text{Exp.}K_{eq} = [H^+] \times\ [HCO_3^-]/[CO_2] = 4.5 \text{ x } 10^{-7}\ \left(\log\ K = -6.35 = -\text{apparent } pK_a\right)$$

While the forward reaction is slow, the reverse of this reaction is catalyzed by the proficient enzyme, carbonic anhydrase (CA), and this reaction is fast and practically diffusion limited. The empirical and experimental values of energetics and kinetics seem to agree.

$$CO_2\ (-386) + OH^-\ (-569.7) \rightarrow HCO_3^-\ (-999.8);\ \Delta_r G° = -44.1\text{kJ/mol};\ \approx K_{eq} = 5.4 \text{ x } 10^7 M^{-1}\ (2q)$$

$$\text{Exp.}K_{eq} = [HCO_3^-]/[CO_2] \times\ [OH^-] = 4.6 \text{ x } 10^7 M^{-1}$$

Rate constants of the forward and reverse reaction are $8.5 \times 10^3\ M^{-1}\,s^{-1}$ and $2 \times 10^{-4}\ s^{-1}$, respectively [68]. The forward reaction may be catalyzed by CA, and this is fast and practically diffusion limited. There is a fair agreement on empirical calculation and experimental equilibrium constant; energetics and kinetics are also in agreement.

$$H_2CO_3\ (-623.1) \rightarrow H^+\ (412.5) + HCO_3^-\ (-999.8);$$
$$\Delta_r G° = 35.8\text{kJ/mol};\ \approx K_{eq} = 5.3 \text{ x } 10^{-7} M\ (2r)$$

$$\text{Exp.}K_{eq} = [H^+] \times\ [HCO_3^-]/[H_2CO_3] = 2 \text{ x } 10^{-4} \text{ OR } \left(\text{empirical theory } 6.0 \text{ x } 10^{-7}\right)$$
$$\left(\log\ K = -3.69 = -\text{true } pK_a\right)$$

The forward reaction's rate constant is $1\times10^{7}\ s^{-1}$ whereas the reverse as $5\times10^{10}\ M^{-1}\ s^{-1}$ [69]. Ultrafast reactions observed both ways. Emperical energetics favors the reverse reaction. The direction of experimental K_{eq} and overall kinetics agrees with energetics.

3.4.1.3 Formation of carbonate ion

We can consider two ways.

$$HCO_3^-\ (-999.8) \rightarrow H^+\ (412.5) + CO_3^{2-}\ (-1347.7);$$
$$\Delta_r G = 65.3 kJ/mol;\ \approx K_{eq} = 3.6 \text{ x } 10^{-12} M\ (3m)$$

$$\text{Exp.}K_{eq} = [H^+] \times [CO_3^{2-}]/[HCO_3^-] = 4.7 \text{ x } 10^{-11}\ (\log\ K = -pK_a = 10.33)$$

$$CO_2\ (-386) + H_2O\ (-237.2) \rightarrow 2H^+\ (412.5) + CO_3^{2-}\ (-1347.7);$$
$$\Delta_r G^\circ = 100.5 kJ/mol;\ \approx K_{eq} = 2.4 \text{ x } 10^{-18} M\ (3n)$$

$$\text{Exp.}K_{eq} = [H^+]^2 \times [H_2CO_3]/[CO_2] \times [H_2O];\ (\log\ K = -17.45)$$

Practically, these are the ultraslow reactions that would have little relevance in physiology. Once again, the energetics and kinetics are in agreement.

In the various steps, it is seen that kinetics aligned with thermodynamic disposition. If α is the fraction of the species of the three dissolved derivatives of CO_2 (H_2CO_3, HCO_3^- and CO_3^{2-}), the profiles of the three species can be traced with respect to pH. With simple chemical equilibrium considerations, it can be clearly seen/inferred that bicarbonate ion is the overwhelming species in the physiological conditions of chloroplast (at pH $\sim$8), with very little carbonic acid or carbonate formation occurring from the direct dissolution or equilibration of gaseous CO_2. The point to note is that the value of the theoretical equilibrium constant for the direct hydration of CO_2 (reaction 1, mediated by CA) is close to zero and the reaction would be highly facile in physiological milieu. The detailed considerations above are presented because they are required to elucidate the interactive chemistry in dynamic/steady-state conditions (wherein a participant of equilibrium like bicarbonate could be actively formed and consumed by metabolic processes) within a plant physiological milieu.

3.4.2 Murburn 1e/2e equilibriums, with bicarbonate as electron donor

Contrary to the classical purview of affinity-based and deterministic rationales of serial electron transfers, murburn proposal is based in understanding physiological reactions in terms of stochastic/statistical events/outcomes resulting from a milieu that contains diverse molecules and ions of various activity/mobility and redox potentials. In this purview, any component can react at any juncture of the interactive electron/moiety transfer equilibrium; provided that they are present in appropriate concentrations and presented in the precise locus, with a favorable orientation. Continuing from the earlier section's discussion, we can see that the classical CA-catalyzed facile equilibrium favored reactions can be represented as:

$$CO_2\ (-386) + OH^-\ (-569.7) \rightarrow HCO_3^-\ (-999.8);\ \Delta_r G^\circ = -44.1 kJ/mol \quad (1)$$

$$H^+\ (412.5) + HCO_3^-\ (-999.8) \rightarrow CO_2\ (-386) + H_2O\ (-237.2);\ = \Delta_r G^\circ = -35.9 kJ/mol \quad (2)$$

$$H_2CO_3\ (-623.1) \rightarrow CO_2\ (-386) + H_2O\ (-237.2);\ \Delta_r G^\circ = -0.1 kJ/mol \quad (3)$$

Therefore, the CA activity would be deemed pH dependent. The first reaction occurs on the alkaline side and leads to the formation of bicarbonate. The second reaction occurs on the acidic side and results in the loss of bicarbonate. Since chloroplast physiology of pH 8 favors bicarbonate formation by CA, effective utilization of the same is a definite theoretical option. The reverse of (3) could equate to (2) in steady-state, and this reaction would be expected to be highly viable kinetically if accompanied by exergonic reactions (as envisaged during the events resulting post photo-activations). After the events of photo-activation and electron donation, we consider the premises wherein an electron can be abstracted from bicarbonate ion or carbonic acid, via murburn photolytic reactions.

$$HCO_3^- \ (-999.8) \rightarrow CO_2\ (-386) + {}^*OH\ (25.8) + e^-\ (-148.53); \Delta_r G^\circ = 491.1 kJ/mol \quad (4)$$

$$H_2CO_3\ (-623.1) \rightarrow CO_2\ (-386) + {}^*OH\ (25.8) + e^-\ (-148.53) + H^+\ (412.5); \ \Delta_r G^\circ = 527 kJ/mol \quad (5)$$

From the above $\Delta_r G^\circ$ values of the equations, we had recently proposed for electron abstractions in transformed scale $\Delta_r G\prime^o{}_{aq}$ [19], are given within the parentheses after equations in bold:

$$ {}^*O_2^-\ (-381.9) \rightarrow O_2\ (16.4) + e^-\ (-148.53); \ \Delta_r G^\circ = 249.8 kJ/mol\ (\Delta_r G\prime^o{}_{aq} = 250 kJ/mol) \quad (6)$$

$$H_2O_2\ (-134.03) \rightarrow {}^*O_2^-(-381.9) + 2H^+\ (412.5) + e^-\ (-148.53); \ \Delta_r G^\circ = 428.6 kJ/mol\ (\Delta_r G\prime^o{}_{aq} = 426 kJ/mol) \quad (7)$$

$$OH^-\ (-569.7) \rightarrow {}^*OH\ (25.8) + e^-\ (-148.53); \ \Delta_r G^\circ = 447 kJ/mol\ (\Delta_r G\prime^o{}_{aq} = 446 kJ/mol) \quad (8)$$

$$H_2O\ (-237.19) \rightarrow {}^*OH\ (25.8) + H^+\ (412.5) + e^-\ (-148.53); \ \Delta_r G^\circ = 527 kJ/mol\ (\Delta_r G\prime^o{}_{aq} = 525 kJ/mol) \quad (9)$$

It can be seen that both values are in good agreement, ratifying our methods. It is forthright to deduce that if the electron can be abstracted from the water molecule (as indicated by isotope analysis of [3]), the involvement of diffusible reactive oxygen species (DROS- like $^*O_2^-$, H_2O_2, $^*OH$, OH^-, $^1O_2$, etc.) in the scheme cannot be negated due to theoretical considerations and experimental observations. Also, bicarbonate ion could serve as a direct source of electrons and oxygen atom for the photosynthetic intermediates, as it is more viable. It can be seen in either case, the electron or oxygen may be indirectly/ultimately sourced from water/hydroxide ion, via the following two considerations A & B.

(A). 1 + 2p = 2r, followed by 2r + (4)

$$(1)\ CO_2 + H_2O \rightarrow H_2CO_3; \Delta_r G^\circ = 0.1 kJ/mol$$

$$(2p)\ H_2CO_3 \rightarrow H^+ + HCO_3^-; \Delta_r G^\circ = 35.8\ kJ/mol$$

$$(2r)\ CO_2 + H_2O \rightarrow H^+ + HCO_3^-; \Delta_r G^\circ = 35.9 kJ/mol$$

$$(4)\ HCO_3^- \rightarrow CO_2 + {}^*OH + e^-; \Delta_r G^\circ = 491.1 kJ/mol$$

$$\text{Totaled}: H_2O \rightarrow {}^*OH + H^+ + e^-; \Delta_r G^\circ = 527 kJ/mol \approx \text{Rxn.}(9)$$

(B). 2q + (4)

$$(2q)\ CO_2 + OH^- \rightarrow HCO_3^-; \Delta_r G^o = -44.1 kJ/mol$$
$$(4)\ HCO_3^- \rightarrow CO_2 + {}^*OH + e^-; \Delta_r G^o = 491.1 kJ/mol$$

$$\text{Totaled}: OH^- \rightarrow {}^*OH + e^-; \Delta_r G^o = 447 kJ/mol \approx \text{Rxn.}(8)$$

Now, any electron-deficient photo/redox system which has the following features (ECS modalities) can avail electron from various species to regenerate their native state, and it can be seen that the value for bicarbonate lies in between the ones of hydroxide ion and water molecule.

$$PS^+ + OH^- \rightarrow PS + {}^*OH \text{ or } \{\Delta_f G^o(PS) — \Delta_f G^o(PS^+) = -596 kJ/mol\} \quad (10)$$

$$PS^+ + HCO_3^- \rightarrow PS + CO_2 + {}^*OH \text{ or } \{\Delta_f G^o(PS) — \Delta_f G^o(PS^+) =\sim -612 kJ/mol\} \quad (11)$$

$$PS^+ + H_2O \rightarrow PS + {}^*OH + H^+ \text{ or } \{\Delta_f G^o(PS) — \Delta_f G^o(PS^+) =\sim -676 kJ/mol\} \quad (12)$$

After electron abstraction and formation of hydroxyl radical, oxygen evolution is a discretized or delocalized process (and not localized at MnComplex alone!). The overall energetics would be as shown below for single oxygen molecule evolution process:

$$4OH^-\ (-569.7) \rightarrow O_2\ (16.4) + 2H_2O\ (-237.19) + 4e^-\ (-148.53); \Delta_r G^o = 1227 kJ/mol \quad (13)$$

$$2HCO_3^-\ (-999.8) \rightarrow O_2\ (16.4) + 2CO_2\ (-386) + 2H^+\ (412.5) + 4e^-\ (-148.53); \Delta_r G^o = 1475 kJ/mol \quad (14)$$

$$2H_2O\ (-237.19) \rightarrow O_2\ (16.4) + 4H^+\ (412.5) + 4e^-\ (-148.53); \Delta_r G^o = 1547 kJ/mol \quad (15)$$

It can be seen from reactions (4), (5), (A) & (B) that all reactions occurring via bicarbonate amount to ~527 kJ/mol or lower energy terms (since heterolysis of water incurs ~79 kJ/mol, which is the difference between reactions A & B), making it a viable option with respect to water (which requires the same energy). Clearly, the energetics is favorable with bicarbonate and this consideration explains the enhanced oxygen evolution (or electron transfer processes) in the presence of this ion! Therefore, reaction (14) is quite viable in physiology. The last two reactions above can be rewritten for starting from a single moiety, to enable comparison with earlier equations given in literature:

$$H^+\ (412.5) + HCO_3^-\ (-999.8) \rightarrow \tfrac{1}{2} O_2\ (16.4) + 2e^-\ (-148.53) + 2H^+(412.5) + CO_2\ (-386); \Delta_r G^o = 737 kJ/mol. \quad (16)$$

$$H_2O\ (-237.1) \rightarrow \tfrac{1}{2} O_2\ (16.4) + 2e^-\ (-148.53) + 2H^+\ (412.5); \Delta_r G^o = 773 kJ/mol \quad (17)$$

We would like to point out that the accurate Gibbs free energy calculations we report here are much higher than the misplaced earlier values reported in literature [14, 70], which give 103.8 and 156.1 kJ/mol respectively for reactions (14) and (15), for the same ½O_2 stoichiometry. The difference is also seen in relative terms (for bicarbonate versus water); our calculations showing a difference of ~36 kJ/mol compared to Dismukes' differential of ~52.3 kJ/mol. We have discussed the source of such errors and disagreements (as also seen in section II, Eq. (1)) in energetics/equilibriums in recent communications [24, 27].

3.4.3 Bicarbonate could also catalyze generic murburn processes

Beside the core logic described above (where bicarbonate gets consumed in the reaction), the enhancement proffered by bicarbonate can also result due to catalytic role of bicarbonate. Bicarbonate is an effective activator of peroxide, giving peroxymonocarbonate ion [71, 72], a process also facilitated by CA. Research conducted in the last decade shows that carbonate/bicarbonate ions (erstwhile taken as innocent buffer participants) can serve as catalytic agents in oxidative reactions [73], potentially aiding several 1e and 2e murburn equilibriums in milieu. Specifically, bicarbonate ion has been known to enhance photolysis of water by aluminum porphyrins, which are known to proceed via 1e mechanism [74]. This observation is directly relatable to the chemistry of chloroplast reactions. Such positive effects (quite similar to how the addition of some ash to common sugar can enhance its burning in air; another example of a non-deterministic oxidative reaction!) could also explain the enhancement of oxygenic photosynthesis by bicarbonate. In essence, this results because of the lowering of activation energy by the catalyst and by virtue of molecule-unbound ion-radical equilibriums' interactive dynamics, leading to facile sinking of electron pairs into products.

3.5 Explaining discretized oxygenesis, NADP reduction, and ADP phosphorylation

In the Z-scheme, two molecules of water must stay bound to the Mn-complex until several rounds of electrons are transferred through the Z-scheme; DROS production is considered futile and physiological aberrations. In the murburn model, the *OH and other DROS like superoxide formed in the mileu could react/collapse with another similar molecule (or dismutate or cross-react) to form 2e stabilized products of hydrogen peroxide, water, and oxygen molecules. A DROS product like peroxide would also react with the originally formed radicals, further propagating the highly spontaneous and fast binary reactions. Some examples are shown below:

$$^{*}OH\ (25.8) + {^{*}OH}\ (25.8) \rightarrow H_2O_2\ (-134.03); \Delta_r G^\circ = -185.6 kJ/mol \tag{18}$$

$$^{*}O_2^-\ (-381.9) + {^{*}O_2^-}\ (-381.9) + 2H^+\ (412.5) \rightarrow H_2O_2\ (-134.03) + O_2\ (16.4); \Delta_r G^\circ = -158.8 kJ/mol \tag{19}$$

$$^{*}OH\ (25.8) + {^{*}O_2^-}\ (-381.9) + H^+\ (412.5) \rightarrow H_2O\ (-237.19) + O_2\ (16.4); \Delta_r G^\circ = -277.4 kJ/mol \tag{20}$$

$$^{*}O_2^-\ (-381.9) + H_2O_2\ (-134.03) + H^+\ (412.5) \rightarrow {^{*}OH} + H_2O + O_2; \Delta_r G^\circ = -91.8 kJ/mol \tag{21}$$

$$^{*}OH\ (25.8) + H_2O_2\ (-134.03) \rightarrow {}^{*}O_2^{-}\ (-381.9) + H^{+}\ (412.5) + H_2O\ (-237.19);\ \Delta_r G^{\circ} = -98.4\,kJ/mol \tag{22}$$

Thereafter, peroxide could also serve as an electron source to the photo-activated electron-discharged photocatalysts in thylakoids (including photosystems or LHC species), thereby giving rise to superoxide radical [eq. (7)]. Such peroxidase/dismutase reactions could be efficiently carried out by an agent like Mn-Complex of PS II or any other heme in milieu could also serve this role or even CA like enzymes. The reactions proposed herein [particularly, such as (7) and (16)] are supported by the demonstration that Mn-substituted CA works as a peroxidase (in assistance with bicarbonate) and the inference made therein that the overall process could involve radical chemistry [75].

After ECS, at the two photosystems (as detailed in **Figure 9**), electrons can be taken up Fd, which aids NAD(P)$^{+}$ reduction and this proposal is supported by Arnon's group works through 1980s [49]. It can now be understood that that at high interfacial area permitted by the stacking of thylakoids and low water activity (practically aprotic conditions), the DROS radicals are stable and effectively drive phosphorylations [76]. Quinols in the memebrane merely aid these processes by serving as 1e/2e pitstops [25]. This consideration explains how/why quinones/quinols serve as donor/acceptors of both PS I and II (as discussed in Section 3.1, **Figure 2**). Besides the Eqs. (17)−(19), oxygenesis could also accompany photosphorylation steps, which are aided by the various membrane protein complexes that bind ADP (**Figure 9**). Therefore,

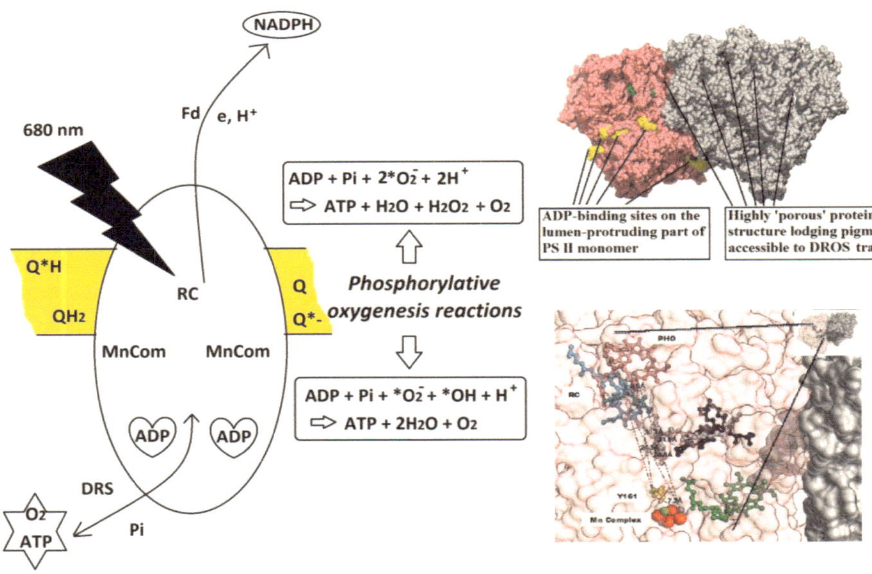

Figure 9.
The murburn explanation for a photosystem (e.g., PS II) and its justification in known structural details: After excitation at 680 nm, RC loses an electron by ECS, which is taken up by ferredoxin, which subsequently reduces NADP, directly or indirectly. The electron-deficient RC can be served by a multitude of species, including DROS like superoxide or peroxide or hydroxide and other species like bicarbonate or water. Quinones in the membrane help stabilize the ECS and also serve as transient pit-stops for the electrons, before they are recycled. The DROS species that could be formed in the meanwhile carries out ADP phosphorylation (as per [19, 22, 26]), aided by the multitude ADP site on the non-membrane portion (right top image). This reaction may also lead to oxygenesis (two equations shown in boxes are thermodynamically facile). The structural constituents (panels on the right) and observed distributions agree well with the murburn model.

oxygenesis is aided by chloroplast-membrane proteins (like PS II/NDH/cytochromes) or soluble proteins (CA/peroxidase) or it could also result due to DROS-cross reactions in milieu. For other evidence, arguments and equations of murburn phosphorylations and NAD(P) reductions, please refer our earlier works [19, 21, 26] (**Figure 9**). The presence of bicarbonate serves as an effective ionic conduit/catalyst/substrate in the 1e and 2e murburn equilibriums occurring in the vicinity of thylakoid membranes.

It can be seen by the analyses of the known structures of Photosystem II that neither TMPD nor PQ (synthetic/natural molecules that can give and receive electrons to/from both photosystems, **Tables 1** and **2**) can reach purported binding sites of PS II (like Mn-complex or non-heme Fe). We do not envisage how even much smaller molecules, such as water or peroxide, could reach the Mn-complex at a steady rate. In this regard, we have shown that redox proteins like peroxidase could abstract electrons via interactive equilibriums in milieu, without the final donor actually getting into the heme active site [77]. Just as water formation was mistakenly considered to happen only at Complex IV (by oxygen staying bound, waiting for 4e and $4H^+$) of the respiratory energetic scheme, oxygen formation/liberation is erroneously perceived to occur only at PS II's Mn-complex (by binding two water molecules, liberating 4e, $4H^+$ and one oxygen molecule). While Complex IV and Mn-complex may be major peroxidase-type murzymes that enable a strong displacement of the equilibrium by effective consumption of accumulated peroxide (forming water and oxygen, respectively), oxygen formation at multiple loci within milieu would also be facile [as shown in the equations above and via Eqs. (17)–(19)]. This consideration also explains the experimental observation of oxygenesis even with 700 nm light, in the Emerson experiment. We have already demonstrated and explained the enhancement of one-electron reactions by several agents like chloride ions in simple peroxidase reactions using murburn concept [77]. The same enhancement effect is also observed in the light reaction of oxygenic photosynthesis [36, 37], thereby confirming the relevance of 1e/2e murburn equilibriums in milieu. The effect can be perceived as akin to the phenomenon wherein distilled water (with miniscule amounts of H^+/OH^- ions) does not conduct electricity but when salts of diverse ions are dissolved in water, it shows better conductivity. Therefore, murburn concept explains the outcomes seen with the non-specific agents in a simpler way, rather than the assumption that all such diverse species have multiple binding sites on different proteins and that such binding could afford allosteric regulatory effects based on conformation changes. Besides being justified in the structure–function correlations and the distributional/experimental facts reported, the bimolecular reactions detailed herein add up thermodynamically (to explain NADP reduction, oxygenesis, thermogenesis and ATP synthesis) [19, 20] and it is also well known that such radical mediated reactions also have high kinetic viability [78, 79].

3.6 Reasoning long-standing observations/conundrums with the murburn model

a. In the murburn model, since electrons can be given by multiple agents and received by diverse species, there is considerable flexibility for probabilistic fecundity. Since all the protein components work independently and in parallel in the murburn model, the Emerson enhancement effect (a few mainstream media perceptions can be seen from the two internet links made by professionals in the field: https://www.maximumyield.com/photosynthesis-maximized/2/924; https://www.youtube.com/watch?v=AJZXFP8ynGA) of oxygenenesis or phosphorylation is explained adequately. Owing to the chemical disposition of the participants (oxygen is a "free to roam and react species"!), the formation of DROS is inevitable and the spontaneous bimolecular oxygenesis reactions discussed above cannot be 'prevented' by preset deterministic mandates that Z-scheme imposes. There exists adequate scope for bicarbonate to effectively serve a constructive role in oxygenesis [say, via (4) and (18) or (14)+(19)].

b. The requirement of the external protons (on the left side of the equations) and their ability to displace the oxygenesis equilibriums [(17)–(19)], reduction of nicotinamide nucleotide (via $NADP^+ + H^+ + 2e^- \rightarrow NADPH$) and phosphorylations (via $ADP + Pi + H^+ \rightarrow ATP + H_2O$ or through one-electron murburn equilibriums leading to superoxide generation in situ) also explain the enhancement of photosynthesis in the Jagendorf experiment [80]. Therefore, the pmf-based CRAS was erroneously assumed to be a valid proposal.

c. CA is copious in chloroplasts and it is a highly efficient enzyme. It catalyzes the spontaneous/facile proton-utilizing formation of CO_2 (the physiological substrate of RUBISCO), the hydroxide-utilizing formation of bicarbonate and the slightly endergonic hydration of CO_2. (the latter two reactions giving murburn substrates for oxidized photosystems/pigments). Therefore, the presence or the preponderance of CA (and CA-like activity of Photosystem II or any other agent thereof) in chloroplasts can be deemed relevant. However, the CO_2 production by CA may not be obligatorily required physiologically in C3 plants [64] because the CA-type activity of Photosystem II may substitute in lieu. Since CA's absence leads to altered pH and DROS dynamics, it is projected that CA could serve as a murzyme (with the zinc atom cycling via a transient one-electron reduced state). Such a mechanism could explain this enzyme's practically diffusion-limited turnovers and the requirement of the metal cofactor. Since CA and Photosystem II are present at high copy numbers in chloroplasts, they have the ability to influence the in situ bicarbonate-involving equilibriums discussed in section 2. In this regard, it is essential to consider the arguments of Warburg [81] and the unexplained data several groups presented thereafter [10, 11, 82, 83]. For the readers' convenience, the major experimental dat profiles of cited above are reproduced from the original sources in Item 3 of the Appendix. The following are direct explanations of and deductions from those works:

1. Our re-interpretations above are supported by the sole figure of Stemler & Radmer's [83] paper in *Science*, dealing with the provision of ^{18}O-labeled bicarbonate to disrupted chloroplasts depleted of bicarbonate/CO_2 and lacking the enzyme CA. (This is shown as panel (A) in Item 3 of Appendix.) They observed that: (a) the immediate/instantaneous formation of heavy-atom labeled dissolved CO_2 upon the provision of labeled bicarbonate, (b) the evolution of unlabeled dissolved CO_2 in milieu occurs in parallel to the evolution of unlabeled oxygen, and (c) though delayed by a few minutes; there is a small amount of label found in oxygen evolved. The first and second observations show that even in CA/CO_2 depleted systems, the HCO_3^- - CO_2 dynamics is instantaneously facile/operative in physiology. If bicarbonate is not involved in photosynthesis, we would not expect any label in evolved CO_2 at all, and this consideration

is violated by observation (c). The low yield of label in evolved oxygen can be explained considering that the maximal concentration of (the labeled oxygen containing) bicarbonate in physiological milieu would only be at a few mM levels, (the unlabeled oxygen containing) water concentration is at 55.6 M (which is a conservative excess of >10^4!). Therefore, labeled bicarbonate could instantly undergo the facile reverse of 2p and 2q. Thereby, the heavy oxygen atom label in bicarbonate would be reduced to a minuscule fraction. However, provision of heavy label in water cannot drive up the label into bicarbonate, owing to equilibrium considerations. Thus, the vast majority of label in evolved oxygen would always be seen as sourced from water. On the other hand, the presence of some label in the evolved oxygen is supportive of bicarbonate's involvement, and cannot be reasoned in any other way! Therefore, it is now deemed inappropriate to argue whether the electrons or oxygen come from water or bicarbonate, as bicarbonate and water are intricately connected via a network of 1e/2e equilibriums in physiology (discussed in sections 2-4).

2. **Figures 1–3** of Radmer-Ollinger's [82] *FEBS Lett.* paper is shown as panel (B) in Item 3 of Appendix. **Figure 1** clearly establishes the controls that practically, little labeled oxygen is seen when the system is presented with labeled oxygen-containing bicarbonate. The authors' data analyses reveal that the amount of labeled oxygen given is a fraction of the total bicarbonate (3 and 32% of oxygen atoms of CO_2 being doubly and singly labeled, respectively!). If we consider radical-rebound reactions being operative (leading to the carbonic anhydrase type outcomes), the kinetic isotope effects would dictate that the heavy atom is not dislodged statistically, making only the lighter atoms of the bicarbonate involve in the reaction. Therefore, the initial timeframes O_2 evolved might not show any label at all, due to low availability at one hand, and low reactivity at the other! The authors correctly reason and correlate with earlier studies that the amplitude of oxygen yields go up with the provision of bicarbonate, an important aspect which is conveniently sidelined by Z-scheme advocates (who just focus only on the negative oxygen-label data!). One crucial finding that everyone misses is that there is significant labeled oxygen evolution in the 2nd flash also, which is accentuated by the addition of bicarbonate. **Figure 2** of their paper is the exploration of oxygen evolution with the provision of labeled and unlabeled oxygen-containing water. The first observation is that the labeled water signals for labeled oxygen was lower by at least eight folds (in comparison to unlabeled oxygen production with unlabeled water), confirming the inference of kinetic isotope effects lowering rates (i.e., radical rebound mechanism being operative). Once again, the larger amplitudes obtained in these traces confirm that oxygen is evolved even after second light flash (equivalent to the first flash of Kok-Joliot experiments, as a priming flash is unaccounted in their protocols!). In the labeled water experiment, this result is inexplicable with the classical Mn-complex centered Kok-Joliot cycle, which would require prior-bound ^{18}O intermediates that must undergo a mechanistically impossible "sequential double-hits" at a non-photo-excitable center [19]. This earlier reported finding conclusively disclaims the classical Kok-Joliot model and provides strong support for the murburn oxygenesis mechanism. **Figure 3** of the same paper shows that bicarbonate significantly enhances this

second flash's yields whereas a control like sodium chloride does not give similar outcomes! Surely, this necessitates that the "double-hit" argument must be jettisoned, as there is no foreseeable way in which bicarbonate could bring this effect. This finding suggests a rapid equilibrium based effect, wherein bicarbonate's presence impacts oxygen evolution, an outcome which is permitted within the murburn radical interactions purview. Further, in the murburn model, the oxygen evolved in the first few pulses could also get consumed internally for competing reactions. The presence of additives, such as the reducible ferricyanide only affects this internal competition, thereby influencing the yields in the second light pulse. The stochastic murburn model allows for multiple such discrete equilibriums (which convincingly explain the outcomes!) whereas the deterministic Z-scheme does not. Also, the quartet periodicity accentuated at the third flash is not a conserved observation (which we had pointed out earlier!), thereby disclaiming the Kok-Joliot cycle.

3. The so-called decisive works in this field by Clausen et al. [10] and Hillier et al. [11] that downplay the roles of bicarbonate do not take into account the observations reported earlier and inferences we made above, and continue to misinterpret the isotope findings, overlooking the major aspects (like kinetic isotope effects and the factual observation of heavy atom trace in oxygen from bicarbonate, albeit at later times!). Hilliers et al. work was done primarily with PS II core-containing membrane fractions, and that too, incorporating CA inhibitors like ethoxyzolamide and high concentration of the oxidizer, ferricyanide. Such a single-turnover experiment has little relevance to physiological steady-state conditions wherein multiple photosystems, LHCs and ample oxygen (and no ferricyanide!) + bicarbonate production mechanism would be present. Showing that oxygen can be produced even in the absence of the physiological ambiance and without labeled isotope containing oxygen is not any evidence to elucidate the actual physiological process (involving bicarbonate)! Even in such reductionist/unrealistic experimental work (which was designed to show that bicarbonate cannot have physiological role; and not designed to investigate the physiological role of bicarbonate!), the authors have themselves admitted that there is significant (although not accounting for the major process!) label in O_2 produced by labeling bicarbonate! **Figure 2** of Clausen et al. paper is presented in Item 3 of Appendix as panel (C). Clearly, it can be seen that even in this experimental work, oxygen evolution is noted after the second flash and once again, the "third peak maximal quartet" paradigm is not reproducible (when compared to a similar experiment in **Figure 2** of Radmer-Ollinger paper). Even in this work, it was seen that labeled oxygen given in water also equilibrates with CO_2 to some extent, and the decay of such heavy labeled CO_2 is attributed to physiological photosynthetic activity. The fact that label is seen in oxygen in later time frames (in Stemler-Radmer paper) and lower yield seen in labeled water (in Radmer-Ollinger paper) shows the kinetic isotope effects involved in radical rebounds. As per the murburn model, CA type activity and multiple other components interact (see discussion below) to contribute the O atom into molecular/gaseous oxygen or electrons in NADPH. Besides

$$HCO_3^- + PS^{+*} \rightarrow CO_2 + {}^*OH + PS$$
$${}^*OH + {}^*OH \rightarrow H_2O_2$$
$${}^*OH + OH^- \rightarrow H_2O + {}^*O^-$$
$${}^*OH + H_2O_2 \rightarrow {}^*O\text{-}O^- + H_2O + H^+ \text{ or } {}^*O\text{-}O^- + H_2O + H^+$$
$${}^*OH + {}^*O\text{-}O^- + H^+ \rightarrow H_2O + O_2$$
$${}^*OH + {}^*H \rightarrow H_2O$$
$$2HCO_3^- + 2{}^*OH + 2H^+ \rightarrow 2CO_2 + H_2O_2 + 2H_2O$$

Figure 10.
*Murburn reaction for bicarbonate activation and steps for loss of heavy atom trace in evolved oxygen. As per the first equation given above, bicarbonate gives the electron-deficient photoactive molecule an electron. (The label O atom is in bold colored font.) This equation is evidenced by Stemler & Radmer paper [83], which shows that as soon as labeled bicarbonate HCO_3^- was presented, labeled CO_2 was produced. This data has to be interpreted in conjunction with the middle image of **Figure 3** of Radmer-Ollinger paper [82], which clearly shows enhanced oxygen peak with second light pulse (upon presenting non-labeled bicarbonate). Further, the sole image in the former paper also shows that oxygen evolved does have label when presented with bicarbonate, albeit the label comes only after some time. This is because of kinetic isotope effects (evident in the labeled water data **Figure 2** of Radmer-Ollinger paper), as ^{18}O is knocked out at approximately an order lesser (~1/8) than the ^{16}O atom. Here, it cannot be argued that the oxygen label is from labeled water formed from the bicarbonate-evolved hydroxyl radical reactions. In which case, it must be accepted that bicarbonate does serve as a source of electrons! In turn, this would lead to the theoretical imperative that it could also be a source of oxygen, albeit to low extents! Therefore, the hydroxyl O-label derived from bicarbonate label must go into water or peroxide or other species. The rest of the murburn equations [26] that follow (after the photosystem mediated electron-abstraction step) explain how the heavy label in $^*OH$ goes into non-O_2 species.*

our theoretical treatments presented herein, our inference of bicarbonate interacting with DROS is supported by findings of Warburg's original observations and several scientists' data [4, 8, 13–16, 81, 83–87]. To sum up, bicarbonate could potentially serve as a catalyst or a source of electrons and/or oxygen in photosynthesis, and the observations noted in connection could be explained by the equations given in **Figure 10**.

3.7 Is the light reaction of photosynthetic plants a deterministic or stochastic process?

The classical sequence or order of the equations/processes occurring via Z-scheme and Mn-complex based oxygen evolution cannot have bicarbonate binding at the WSC/OEC and the PS II must bind bicarbonate elsewhere, like the non-heme iron center of PS II [88]. A study of the Mncomplex containing extra-membrane region does not show major channels connecting to the exterior bulk phase. It is not clear how water molecules could channel into the Mn-center (and likewise, the larger bicarbonate ion would find it even more difficult!). The large bulbous protrusions of the PS II and porous/cavity-ridden nature of the protein at its various loci are inexplicable in the classical purview (**Figures 4** and **9**).

The classical perspective sticks to affinity-driven binding-based causatives alone and since it sees oxygen evolution occurring only at PS II's Mn-complex, researchers have stuck to probing the effects of bicarbonate resulting only from its binding to PS II. Since the non-heme Fe center is not a "route" charted out in the Z-scheme, the outcome can only attributed to (non-traceable or unaccountable) allosteric effects. Regardless, assuming that bicarbonate can and does bind anywhere at PS II (as perceived above by some researchers who tried to explain the bicarbonate enhancement within an overzealous extension of the discredited Z-scheme), the following set of equations formed the basis for the interpretations. That is, bicarbonate splitting gave $4e^-$ and 2 CO_2 molecules, which could be recycled via CA type activity. It can be seen that the sum total energetic yield of this bicarbonate mediated process is equal to simple water splitting in this classical scheme too (refer earlier equation xv), as given below.

$$2HCO_3^- + 2H^+ \rightarrow O_2 + 2CO_2 + 4H^+ + 4e^-$$
$$2CO_2 + 2H_2O \rightarrow 2HCO_3^- + 2H^+$$

$$2H_2O + 4\,h\nu \rightarrow O_2 + 4H^+ + 4e^- \; (\Delta_r G^\circ = 1547 \text{ kJ/mol–}[4\,h\nu \approx 704 \text{ kJ/mol}] = 843 \text{ kJ/mol}) \qquad (23)$$

The other half-reaction occurring at a disconnected locus of PS II would be:

$$2Q + 4H^+ + 4e^- \rightarrow 2QH_2 \; (\Delta_r G^\circ = -1229.34 \text{ kJ/mol}) \qquad (24)$$

While this reaction's mechanistic approach could perhaps enable the justification or the formation of labeled oxygen atoms with the provision of labeled bicarbonate or water, it cannot explain the enhancement of oxygen yields with bicarbonate. Also, it cannot reason the reduction of $NADP^+$ or phosphorylation of ADP independently by this complex. The overall requirement at PS II is 1547 kJ/mol and this is apparently offset by the 704 kJ/mol contribution of four photons and 1229 kJ/mol derived from two quinone molecules' reduction (exceeding the required amount by a value of ~386 kJ/mol). **Figure 4** shows the spatio-temporal overview of the processes that transpires at PS II, as per Z-scheme. It can be seen that the yield obtained for the temporally and spatially disconnected quinone reduction (which completes at 10^{-2} seconds) occurring nanometers away from both the RC (where the energy or electron transfer events occur at 10^{-12} to 10^{-5} second time-frames) and OEC/WSC (where the purported water-lysis occurs in 10^{-3} seconds) cannot be coupled with any known mechanism. This makes the overall process unviable via all considerations (thermodynamic, kinetic, mechanistic and probabilistic). That is, since there is nothing called "half a molecule of O_2" in reality, Z-cheme dictates that only after two $NADP^+$ molecules are completely reduced at the end of multiple ETC cycles can a molecule of oxygen evolve at OEC. Therefore, for one molecule of oxygen to be released by the WSC, PS II must make two QH_2 molecules. The second molecule of Q cannot bind until the first formed QH_2 detaches, and each such quinone must deterministically home its way to QBC. These considerations mean that with 1 $h\nu$: 1e stoichiometry, PS II should orchestrate multiple events at different loci independently! It should split one O-H bond (of ~460 kJ/mol) releasing one electron (± proton), using the insufficient energy provided by a photon of 176 kJ/mol (assuming absolutely efficient energy conservation/coupling between RC and WSC). Further, if we consider that the proton formation/release is not energetically favored and that WSC must have deterministic proton relay networks to enable the reduction of quinone at its binding site (QBC), the ETC-CRAS mechanism appears improbable. (Then again, the membrane is also supposed to be impermeable to proton fluxes, at the same time, to build *pmf*!) Such deterministic perceptions do not limit the stochastic scheme of murburn model, which has ion-radical equilibriums. For greater insights, please refer our conclusive discussions discrediting the various aspects of

"Kok-Joliot cycle–Z-scheme–CRAS" explicatory paradigm presented in our recent publications [17, 19, 21, 23, 25, 26, 58]. While quinols are practically immobile in the reaction time frames and cannot move deterministically in membranes, DRS like superoxide can freely move, relaying electrons. Also, DROS formed in milieu can attack ADP bound to the protein or phosphate found in milieu, thus aiding photo-phosphorylation. For the details of murburn model of NADP reduction and ADP phosphorylation, please refer the murburn precepts paper [19].

In the murburn purview, we can envisage the following set of reactions for PS II + LHC:

$$n\ \mathrm{LHC}/n\ \mathrm{PS} + n\ \mathrm{DS} + n\ hv \rightarrow n\ \mathrm{LHC}^{+*}/n\ \mathrm{PS}^{+*} + n\ \mathrm{DRS}^{-*}\ (\text{e.g. } {}^{*}\mathrm{H}, \mathrm{Mg}^{+}, {}^{*}\mathrm{O_2}^{-}) \tag{25}$$

Under such high potentials generated by multiple membrne-embedded species donating electrons and these electrostatics stabilized transidently via effective charge separation afforded by the various redox centers of PS II, the membrane becomes highly positively charged. Meanwhile, protons keep coming in through the membrane at millisecond timescales and DRS (like superoxide) can take up these protons and undergo dismutations to give peroxide and other DRS. This peroxide can be easily used by Mn-complex to further liberate electrons/O_2 and DRS. Furthermore, the stochastic interactive networks of "membrane complexes and soluble proteins (like Fd/PC) + DRS + water + hydroxide ion + bicarbonate ion + ADP/Pi + NAD(P)" thereafter can also liberate oxygen and make the other products (NADPH and ATP) in the vicinity of PS II. Further, the large lumenal protrusions bearing ADP sites and the distribution of Fd/PC in both stroma/ lumen and the findings of Arnon (reduction of NADP even at PS II) support the murburn model of ADP phosphorylation and NADP reduction. In reductionist systems lacking CA or anoxic initial conditions also, oxygen formation would be possible wherein PS II could abstract an electron directly from a starting species like hydroxide ion. Such murburn processes would also not have energetic or kinetic or mechanistic or probabilistic limitations, as imposed by the erstwhile explicatory paradigm. In such a purview, other reactions of bicarbonate (an example is given below) would also be highly feasible, enabling rapid loss of isotope traces:

$$2\mathrm{HCO_3}^- + 2\,{}^{*}\mathrm{OH} \rightarrow 2\,{}^{*}\mathrm{CO_3}^- + 2\mathrm{H_2O}$$
$$2\,{}^{*}\mathrm{CO_3}^- + 2\mathrm{H}^+ \rightarrow 2\mathrm{CO_2} + \mathrm{H_2O_2}$$

$$2\mathrm{HCO_3}^-\ (-999.8) + 2\,{}^{*}\mathrm{OH}\ (25.8) + 2\mathrm{H}^+\ (412.5) \rightarrow 2\mathrm{CO_2}\ (-386) + \mathrm{H_2O_2}\ (-134.03) + 2\mathrm{H_2O}\ (-237.19); \Delta_r G^\circ = -257.41\ \mathrm{kJ/mol} \tag{26}$$

$$2\mathrm{H_2CO_3}\ (-623.1) + 2\,{}^{*}\mathrm{OH}\ (25.8) \rightarrow 2\mathrm{CO_2}\ (-386) + \mathrm{H_2O_2}\ (-134.03) + 2\mathrm{H_2O}\ (-237.19); \Delta_r G^\circ = -185.81\ \mathrm{kJ/mol} \tag{27}$$

The above "simple chemical engine" (SCE) murburn model for chloroplast resting on "effective charge separation" (ECS) principles is also justified by the underlying stochastic mechanistic principles, as observed in phosphorylation chemistry. In mid-twentieth century, Mildred Cohn had done pioneering experiments with oxygen-labeling in mitochondrial physiological phosphorylations [89, 90]. She had found that oxygen atom labels incorporated into phosphate or ATP were too quick and too numerous to be accounted for by classical enzyme reactions like substrate level phosphorylations. Reinterpreted in the current awareness, this experiment had originally showed that it was the DRS-activity that resulted in the physiological interactive equilibriums. The pioneer's original findings need to be understood and exalted, to really appreciate the dynamics of DR(O)S in bioenergetics. It can be clearly seen that only bimolecular fast radical reactions (as proposed in the current manuscript, via the murburn model) can afford meaningful explanations to the effects observed in bioenergetic chemistry.

4. Summation

By definition, since the Z-scheme sees electron flow from water to NADP (via a defined set of redox centers present on various proteins/biomolecules) as the chartered route, bicarbonate's involvement cannot be explained in the classical purview. This is because each set of electron transfer reaction in Z-scheme is governed by affinity-based logic and tuned to evolutionary perfection. Also, if electrons came in and went out at multiple entry and exit points, there is no point in having the terminology and consideration of a Z-scheme model anyway. In other words, the mandate of the Z-scheme cannot accommodate multiple e-donors or acceptors within physiology. Therefore, its advocates were disinclined to consider data/evidence that violates the ordered sequence of processes that they perceive to be physiologically operative. This was the primary reason that most classical researchers were reluctant to accept the effect of bicarbonate ions (by say, binding on to the non-heme Fe of PS II), in the first place.

We would reiterate that it is inappropriate to view any physiological redox reactions (incorporating multiple redox-active components within milieu) in a deterministic perspective. Within the murburn purview, a given component could have multiple interactive roles, as expected in a probabilistic scheme. This statement is particularly valid when evolutionary mandates may not dictate topographical identification/demarcation of substrates or when the molecules are too small to enable such topological differentiations. We demonstrated this fact with simple reductionist systems, wherein it was shown that classical treatments are inadequate to capture the outcomes observed [77, 91]. It is an evidence driven conclusion that photosynthetic electron transfer processes are not affinity-driven [17, 19, 21, 22, 35, 41] and the overall event is one-electron punctuated processes with a two-electron endings (akin to a paragraph formation using commas and periods) [18]. Murburn model is a stochastic scheme, with no "prescribed route" for reaction outcomes (although preferred routes may exist). The murburn precepts only necessitate an efficient charge separation process at the photosystem so that the photo-ejected electron does not get reabsorbed at the same electron-discharged chlorophyll. This perspective explains the random distribution of the multitudes of pigments and redox-active components in chloroplasts. The murburn purview also supports the evidence-backed proposal that several ions, such as bicarbonate or chloride, may enhance reaction outcomes in photosynthesis [7, 36, 37]. We have demonstrated herein that this effect is owing to thermodynamic and kinetic reasons, wherein such species play favorable roles in murburn ion-radical equilibriums, and they need not directly bind to any particular locus of the photosystems. In this context, it is inappropriate to continue to consider DR

(O)S as a mere manifestation of pathophysiology. Evolution should have done away with the multitudes of 1e centers if DRS were merely deleterious. We also know that all molecular species have contextual relevance. That is: the right amounts of a given DRS at the right places could be useful to physiology. Wrong amounts of an unwanted DRS at the inappropriate place would surely be unproductive to life order. The inferences presented herein and the relevance of DRS in routine physiology is further corroborated by the ubiquitous nature and pan-systemic appeal of murburn concept. The murburn explanation for photo-transduction in vision [92] and the relevance of murzyme functioning for explaining observations and theoretical premises in diverse phospholipid-laden systems of endoplasmic reticula (microsomes), mitochondria, rod cell outer segments, etc. [76] are supportive testaments to this claim. Given that this macroscopic awareness affords a more comprehensive and tangible picture than deeming localized oxygen evolution at Mn-Complex alone, the findings on structures of PS II need to be re-interpreted. Although we could not find water-access modalities to the Mn-center in our analyses, some reports from other groups demarcate several channels [93]. If this were in fact the case, how is it physically possible to restrict two specific water molecules at the active complex, simultaneously preventing the formation/release of DRS at/from PS II [94]? This query would still remain inexplicable in the classical perspective, seeking more and more intangible apologetics of the classical view. Given the fact that the Z-scheme-KokJoliot cycle modality of photosynthesis (which was incorporated in the mega ETC-CRAS paradigm of bioenergetics) has been theoretically dissected and demonstrated to be infeasible, and also the premise that the classical perspectives have been experimentally nullified by multitudes of observations [18–20, 22, 25], the classical perspective should be deemed a redundant aspect of the past. In the least, an honest discussion should be initiated between the workers, mediated by biochemists and biophysicists that do not pose conflict of interests. Else, the future generations would continue to be taught redundant ideas [95, 96], limiting critical thought and potentially waylaying the outlooks availing more efficient green methodologies for harnessing the potential of photosynthesis.

To ratify the murburn chemistry in oxygenic photosynthesis, one of the easiest ways is to add bicarbonate at various concentration ranges and note the outcomes on oxygenensis, nicotinamide reduction and ADP phosphorylation. Since these processes are intricately woven into murburn chemistry, we predict that the stoichiometry of observed would be uncertain/variable, and might show unusual concentration-based effects. Also, this effect is not expected only at PS II, as the outcome is based on multiple reaction equilibriums in milieu (and not necessarily binding to a particular site on a protein!). At this level of awareness, although it is evident that the photosynthetic chemistry would show a variable/uncertain stoichiometry owing to murburn operational principles, the overall process can now be minimally represented as: $2AH/2H_2A + nh\nu \rightarrow A_2 + 2H^+/4H^+ + 2e^-/4e^-$ for simple substrate precursor molecules. Since the murburn model proposes that proton consumption is more favored in bioenergetic routines, the latter route appears more probable. The insights herein could be availed for designing murburn-based photosynthetic (bio)reactors.

Acknowledgements

The work was powered by Satyamjayatu: The Science & Ethics Foundation.

Author contributions

KMM conceived the problems, solutions and wrote the first draft of the paper. YW presented crucial arguments and literature, pointed out key theoretical aspects involved and helped in rewriting the paper. NMB calculated the free energy of formation of carbon dioxide in water and cross-checked the equations. AM provided inputs and helped in shaping the paper. AP made some images of photosystems and analyzed structures.

Funding statement

The work was not funded by any specific agency/body.

Conflict statement

The authors have no conflict of interests to declare.

Data statement

All data needed for the perusal of this document are presented within the manuscript text/supplementary information or duly cited therein.

Abbreviations

ANBS	adenosine nucleotide binding site
CA	carbonic anhydrase
CRAS	chemiosmotic rotary ATP synthesis
CYP	cytochrome P450
CPR	cytochrome P450 reductase
DR(O)S	diffusible reactive (oxygen) species
ECS	effective charge separation
ETC	electron transport chain
Fd	ferredoxin
FNR	ferredoxin-NADP reductase
KJC	Kok-Joliot cycle, K-J cycle
LHC	light harvesting complex
OxPhos	oxidative phosphorylation
PC	plastocyanin
PhotoPhos	photophosphorylation
pmf	proton motive force
PoPs	pumped out protons
PS	photosystem
Q/QH_2	quinones/quinols
RC-Chl	reaction center chlorophyll
TMP	trans-membrane potential

WSC water splitting complex (also called OEC, oxygen evolving complex or Mn-complex)

A. Appendix

Item 1: Survey of electron acceptors from photosystems
Item 2: Thermodynamics & Kinetics

K_{eq} and log K values are from Stryer's textbook or from the following web sources: (http://ion.chem.usu.edu/~sbialkow/Classes/3650/Carbonate/Carbonic%20Acid.html) (https://lawr.ucdavis.edu/classes/ssc102/Section5.pdf).

No.	Name	Molecular Formula	Mass (g/mol)	From PSI/PSII	MV (cm^3)	PSA ($Å^2$)	log P	H-Bonds	Rotating bonds
1	Tetrazolium blue	$C_{40}H_{32}Cl_2N_8O_2$	727.6	I	na	87.6	na	8.0	9
2	DCPIP	$C_{12}H_7NCl_2O_2$	268.0	I	186.7	50	1.91	3.1	1
3	Benzoquinone	$C_6H_4O_2$	108.0	I/II	86.0	34.1	0.26	2.0	0
4	Flavin mononucleotide	$C_{17}H_{22}N_4NaO_{10}P$	496.3	I/II	na	214	−1.7	10.3	6
5	Ferrioxalate	C_6FeO_{12}	319.9	I	na	na	na	na	na
6	$HgCl_2$	HgCl2	271.4	I/II	na	0	na	0.0	0
7	Methyl red	$C_{15}H_{15}N_3O_2$	269.3	I	230.1	52	4.91	5.1	4
8	Paraquat	$C_{12}H_{14}Cl_2N_2$	257.1	I	na	7.8	1.7	0.0	1
9	Ferricyanide	C_6FeN_6	211.9	I/II	na	143	na	12.0	0
10	Anthraquinone	$C_{14}H_8O_2$	208.2	I	159.1	34	3.38	2.0	0
11	Diquat	$C_{12}H_{12}N_2$	184.2	I	na	8	−4.7	2.0	0
12	Napthoquinone	$C_{10}H_6O_2$	158.2	I	122.6	34	1.79	2.0	0
13	Silicotungstate	$H_{72}Na_4O_{40}SiW_{12}$	2970.1	II	na	657	na	40.4	8

Table A1.
A compilation of electron acceptors from the two photosystems.

For calculation of thermodynamic constants, the following steps were used.

1. CO_{2gas}

   In the gas phase in a standard system $\Delta_f(CO_{2gas}) = -394.4$ kJ/mol

2. CO_{2aq}

   When CO_2 is dissolved in water, there are two types of particles in solution: CO_{2aq} and H_2CO_3. H_2CO_3 occurs in the equilibrium reaction

$$CO_{2aq} + H_2O = H_2CO_3 \tag{28}$$

Thus, CO_2 in the gas phase is in equilibrium simultaneously with two particles

$$CO_{2aq} = CO_{2aq} + H_2CO_3 \tag{29}$$

Solubility constant $K_H = 3{,}4 \cdot 10^{-2}$ and $\Delta_r G = 8.35$ kJ/mol. The concentration of H_2CO_3 is low and therefore the value $\Delta_r G = 8.35$ kJ/mol refers to the dissolution of CO_2. Hence,

$$\Delta_f\left(CO_{2aq}\right) = -394.4 + 8.35 = -386 \text{ kJ/mol}.$$

3. H_2CO_3

It is generally accepted that $\Delta_r G(\text{reaktion } 1) \approx 0$

Then $\Delta_f(H_2CO_3) = \Delta_r G(1) + \Delta_f\left(CO_{2aq}\right) + \Delta_f(H_2O) = 0 - 386 - 237.1 = -623.1$ kJ/mol

4. HCO_3^-

HCO_3^- arises from acid dissociation

$$H_2CO_3 = HCO_3^- + H^+$$

The dissociation constant is $4{,}47 \cdot 10^{-7}$, and $\Delta_r G^o = 36.78$ kJ/mol.

Hence,

$\Delta_f G^o\left(HCO_3^-\right) = \Delta_r G^o + \Delta_f G^o(H_2CO_3) – \Delta_f G^o(H^+) = 36.78–623.1–412.5 = -999.82$

5. Dissociation of HCO_3^- is characterized by the constant $4.68 \cdot 10^{-11} = 10^{-10.33}$.

Hence

$$\Delta_r G^o = 64.56 \text{ kJ/mol}$$

$$HCO_3^-\ (-999.8) = H^+(412.5) + CO_3^{2-}\ \ \Delta_r G^o = 64.56 \text{ kJ/mol}$$

Hence,

$$\Delta_f G^o\left(CO_3^{2-}\right) = -1347.7 \text{ kJ/mol}$$

A compilation of standard thermodynamic parameters of particles (arising from the dissolution of CO_2 in water) is given in **Table A2**.

Item 3: Earlier published data on bicarbonate/water labeling and oxygenesis

[Panel (A): Stemler A, Radmer R. [83]. Panel (B): Radmer R, Ollinger O. [82]. Panel (C): Clausen et al. [10]. The resolution is deliberately reduced so as to avoid copyright issues. The readers may peruse the pertinent articles and study the areas highlighted in yellow.

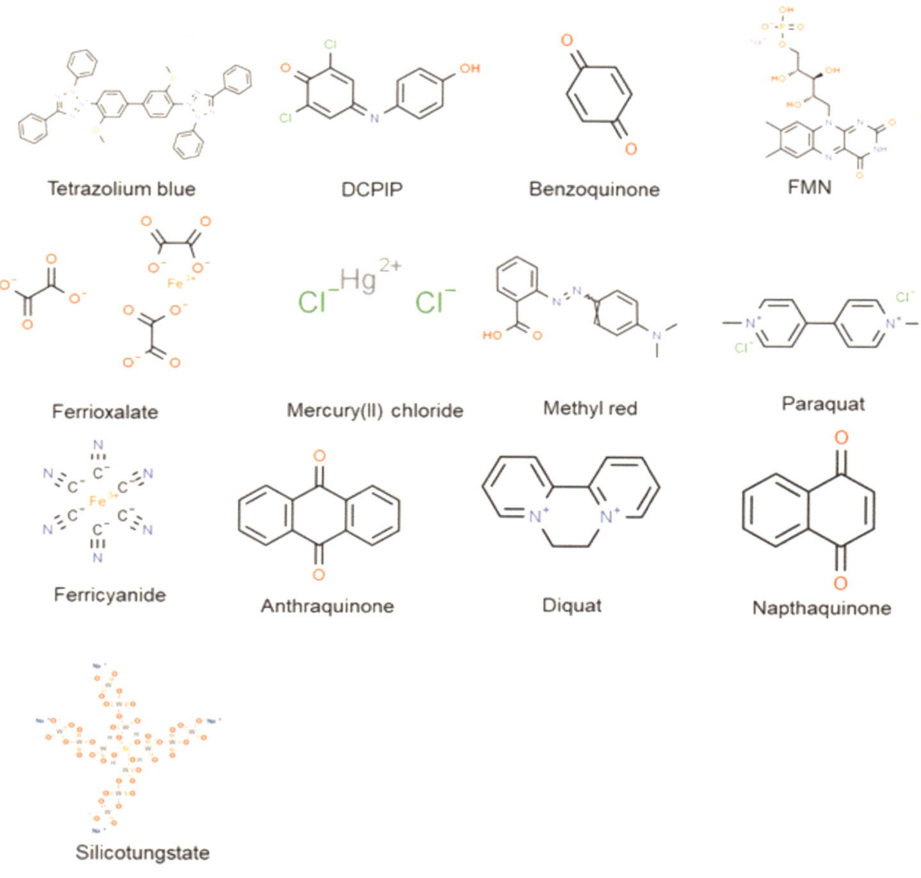

Figure A1.
Structures of various electron acceptors listed in ***Table A1****.*

Molecule	$\Delta_f H^o$, kJ/mol	$\Delta_f G^o$, kJ/mol
H_2O_{aq}	−285.8	−237.1
CO_{2gas}	−393.5	−394.4
$CO_{2,aq}$	−413.8	−386.0
H_2CO_3	−699.2	−623.1
HCO_3^-	−1078.3	−999.8
CO_3^{2-}	−1449.6	−1347.7

Table A2.
Standard thermodynamic properties of species considered in this study.

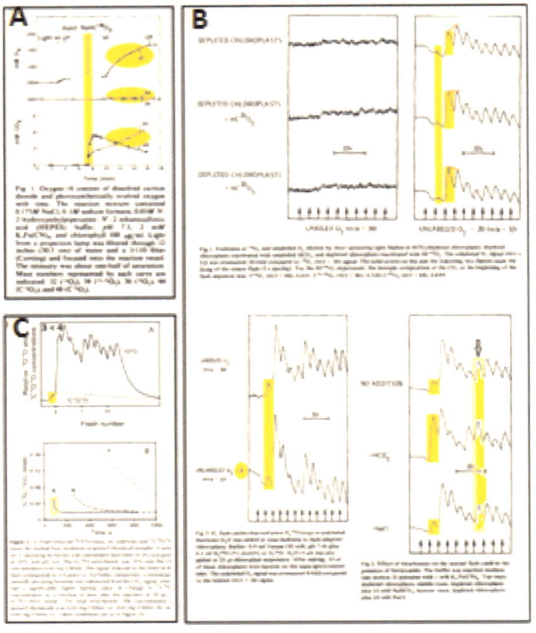

Author details

Kelath Murali Manoj[1]*, Nikolai Bazhin[2], Abhinav Parashar[1], Afsal Manekkathodi[3] and Yanyou Wu[4]*

1 Satyamjayatu: The Science and Ethics Foundation, Palakkad, India

2 Institute of Chemical Kinetics and Combustion, Russian Academy of Sciences, Novosibirsk, Russia

3 Department of Physics, Sultan Qaboos University, Al-khod, Sultanate of Oman

4 State Key Laboratory of Environmental Geochemistry, Institute of Geochemistry, Chinese Academy of Sciences, Guiyang, People's Republic of China

*Address all correspondence to: murman@satyamjayatu.com, a.manekkathodi@squ.edu.om and wuyanyou@mail.gyig.ac.cn

IntechOpen

© 2022 The Author(s). Licensee IntechOpen. This chapter is distributed under the terms of the Creative Commons Attribution License (http://creativecommons.org/licenses/by/3.0), which permits unrestricted use, distribution, and reproduction in any medium, provided the original work is properly cited.

References

[1] Nowicka B, Kruk J. Powered by light: Phototrophy and photosynthesis in prokaryotes and its evolution. Microbiological Research. 2016;**186–187**: 99-118

[2] Whitmarsh J, Govindjee G. The Photosynthetic Process. In: Singhal GS, Renger G, Sopory SK, Irrgang K-D, Govindjee, editors. Concepts in Photobiology: Photosynthesis and Photomorphogenesis. New Delhi: Narosa Publishers and Dordrecht: Kluwer Academic; 1995. pp. 11-51. Available from: https://www.life.illinois.edu/govindjee/paper/gov.html

[3] Ruben S, Randall M, Kamen M, Hyde JL. Heavy oxygen (18O) as a tracer in the study of photosynthesis. Journal of the American Chemical Society. 1941;**63**: 877-879

[4] Warburg O, Krippahl G. Hill-Reaktionen [Hill reactions]. Zeitschrift für Naturforschung. Teil B. 1958;**13B**(8): 509-514

[5] Baranov SV, Ananyev GM, Klimov VV, Dismukes GC. Bicarbonate accelerates assembly of the inorganic core of the water-oxidizing complex in manganese-depleted photosystem II: A proposed biogeochemical role for atmospheric carbon dioxide in oxygenic photosynthesis. Biochemistry. 2000;**39** (20):6060-6065

[6] van Rensen JJS. Role of bicarbonate at the acceptor side of photosystem II. Photosynthesis Research. 2002;**73**:185-192

[7] Stemler AJ, Govindjee. Bicarbonate ion as a critical factor in photosynthetic oxygen evolution. Plant Physiology. 1973;**52**:119-123

[8] Stemler AJ. The bicarbonate effect, oxygen evolution and the shadow of Otto Warburg. Photosynthesis Research. 2002;**73**:177-183

[9] Aoyama C, Suzuki H, Sugiura M, Noguchi T. Flash-induced FTIR difference spectroscopy shows no evidence for the structural coupling of bicarbonate to the oxygen-evolving Mn cluster in photosystem II. Biochemistry. 2008;**47**:2760-2765

[10] Clausen J, Beckmann K, Junge W, Messinger J. Evidence that bicarbonate is not the substrate in photosynthetic oxygen evolution. Plant Physiology. 2005;**139**(3):1444-1450

[11] Hillier W, McConnell I, Badger MR, Boussac A, Klimov VV, Dismukes GC, et al. Quantitative assessment of intrinsic carbonic anhydrase activity and the capacity for bicarbonate oxidation in photosystem II. Biochemistry. 2006;**45** (7):2094-2102

[12] Ulas G, Olack G, Brudvig GW. Evidence against bicarbonate bound in the O_2-evolving complex of Photosystem II. Biochemistry. 2008;**47**(10):3073-3075

[13] Shevela D, Eaton-Rye JJ, Shen JR, Govindjee G. Photosystem II and the unique role of bicarbonate: A historical perspective. Biochimica et Biophysica Acta (BBA) - Bioenergetics. 2012;**1817**: 1134-1151

[14] Wu Y. Is bicarbonate directly used as substrate to participate in photosynthetic oxygen evolution. Acta Geochim. 2021a; **40**:650-658. DOI: 10.1007/s11631-021-00484-0

[15] Wu Y. Bicarbonate use and carbon dioxide concentrating mechanisms in photosynthetic organisms. Acta Geochim. 2021b. DOI: 10.1007/s11631-021-00488-w

[16] Wu YY, Li HT, Xie TX. The regulation on carbon source and carbon sequestration by microalgal carbonic anhydrase, Biogeochemical Action of Microalgal Carbonic Anhydrase. Beijing: Science Press; 2015. pp. 76-111

[17] Gideon DA, Nirusimhan V, Manoj KM. Are plastocyanin and ferredoxin specific electron carriers or generic redox capacitors? Classical and murburn perspectives on two photosynthetic proteins. Journal of Biomolecular Structure and Dynamics. 2022;**40**(5): 1995-2009. DOI: 10.1080/07391102.2020.1835715

[18] Manoj KM, Bazhin N, Manekkathodi A, Wu Y. In: Ameen S et al., editors. Murburn Model of Photosynthesis: Effect of Additives like Chloride and Bicarbonate, Chlorophylls. London: Intech Open; 2022b. DOI: 10.5772/intechopen.103132

[19] Manoj KM, Bazhin NM, Jacob VD, Parashar A, Gideon DA, Manekkathodi A. Structure-function correlations and system dynamics in oxygenic photosynthesis: Classical perspectives and murburn precepts. Journal of Biomolecular Structure & Dynamics. 2021a. DOI: 10.1080/07391102.2021.1953606

[20] Manoj KM, Bazhin NM, Tamagawa H, Jaeken L, Parashar A. The physiological role of complex V in ATP synthesis: Murzyme functioning is viable whereas rotary conformation change model is untenable. Journal of Biomolecular Structure and Dynamics. 2022a. DOI: 10.1080/07391102.2022.2060307

[21] Manoj KM, Gideon DA, Parashar A. What is the role of lipid membrane-embedded quinones in mitochondria and chloroplasts? Chemiosmotic Q-cycle versus murburn reaction perspective. Cell Biochemistry and Biophysics. 2020b;**79**:3-10

[22] Manoj KM, Jacob VD, Parashar A, Gideon DA, Manekkathodi A. Validating the predictions of murburn model for oxygenic photosynthesis: Analyses of ligand binding to protein complexes and cross-system comparisons. Journal of Biomolecular Structure & Dynamics. 2021b. DOI: 10.1080/07391102.2021.1953607

[23] Manoj KM, Manekkathodi A. Light's interaction with pigments in chloroplasts: The murburn perspective. J. Photochem. Photobiol. 2021;**5**:100015

[24] Manoj KM, Gideon DA, Jaeken L. Why do cells need oxygen? Insights from mitochondrial composition and function. Cell Biology International. 2021c;**46**(3): 344-358. DOI: 10.1002/cbin.11746

[25] Manoj KM, Gideon DA, Jaeken L. Interaction of membrane-embedded cytochrome b -complexes with quinols: Classical Q-cycle and murburn model. Cell Biochemistry and Function. 2021d; **40**(2):118-126. DOI: 10.1002/CBF.3682

[26] Manoj KM, Bazhin NM. Murburn precepts of aerobic respiration and homeostasis. Progress in Biophysics and Molecular Biology. 2021;**167**:104-120. DOI: 10.1016/j.pbiomolbio.2021.05.010

[27] Manoj KM et al. Murburn precepts for lactic-acidosis, Cori cycle, and Warburg effect: Interactive dynamics of dehydrogenases, protons, and oxygen. Journal of Cellular Physiology. 2021e;**237**(3):1902-1922. DOI: 10.1002/jcp.30661

[28] Suga M, Akita F, Yamashita K, Nakajima Y, Ueno G, Li H, et al. An oxyl/oxo mechanism for oxygen-oxygen coupling in PSII revealed by an x-ray free-electron laser. Science. 2019;**366**

(6463):334-338. DOI: 10.1126/science.aax6998

[29] Pettersen EF, Goddard TD, Huang CC, Couch GS, Greenblatt DM, Meng EC, et al. UCSF Chimera—A visualization system for exploratory research and analysis. Journal of Computational Chemistry. 2004;**25**(13): 1605-1612

[30] Schrödinger L, DeLano W. PyMOL. 2020. Retrieved from: http://www.pymol.org/pymol

[31] Yu J, Zhou Y, Tanaka I, Yao M. Roll: A new algorithm for the detection of protein pockets and cavities with a rolling probe sphere. Bioinformatics. 2010;**26**(1):46-52

[32] Marcus RA. On the theory of oxidation-reduction reactions involving electron transfer I. The Journal of Chemical Physics. 1956;**24**:966-978

[33] Mehler AH. Studies on reactions of illuminated chloroplasts. I. Mechanisms of the reduction of oxygen and other Hill reagents. Archives of Biochemistry and Biophysics. 1951;**33**:65-77

[34] Hauska G, Oettmeier W, Reimer S, Trebst A. Shuttles of artificial electron donors for photosystem I across the thylakoid membrane. Z Naturforsch C Biosci. 1975;**30**(1):37-45. DOI: 10.1515/znc-1975-1-209

[35] Hauska G. Artificial acceptors and donors. In: Pirson A, Zimmermann MH. editors. Vol. 5. Photosynthetic Electron Transport and Photophosphorylation. Berlin-Heidelberg-New York: Springer; 1977. pp. 253-265

[36] Izawa S, Heath RL, Hind G. The role of chloride ion in photosynthesis. 3. The effect of artificial electron donors upon electron transport. Biochimica et Biophysica Acta. 1969;**180**(2):388-398. DOI: 10.1016/0005-2728(69)90123-6

[37] Kaňa R, Govindjee. Role of ions in the regulation of light-harvesting. Frontiers in Plant Science. 2016;**7**:1849. DOI: 10.3389/fpls.2016.01849

[38] Kelley PM, Izawa S. The role of chloride ion in photosystem II. I. Effects of chloride ion on photosystem II electron transport and on hydroxylamine inhibition. Biochimica et Biophysica Acta. 1978;**502**(2):198-210. DOI: 10.1016/0005-2728(78)90042-7

[39] Magnuson A. Electron donor systems in natural and artificial photosynthesis. [Doctoral Thesis]. Lund University. 1998. ISBN: 91-628-3149-6

[40] Maslenkova A, Zeilanov Y. Effect of some artificial electron donors and acceptors on the functioning of the photosynthetic oxygen evolving system. Bulgarian Journal of Plant Physiology. 1995;**21**:3-11

[41] Tschortner J, Lai B, Kromer JO. Biophotovoltaics: Green power generation from sunlight and water. Frontiers in Microbiology. 2019;**10**:866

[42] Kühlbrandt W. Structure and function of bacterial light-harvesting complexes. Structure. 1995;**3**:521-525

[43] Croce R, van Amerongen H. Light harvesting in oxygenic photosynthesis: Structural biology meets spectroscopy. Science. 2020;**369**:eaay2058

[44] Arnon DI, Allen MB, Whatley FR. Photosynthesis by isolated chloroplasts. Nature. 1954a;**174**:394-396

[45] Arnon DI, Whatley FR, Allen MB. Photosynthesis by isolated chloroplasts. II. Photosynthetic phosphorylation, the

conversion of light into phosphate bond energy. Journal of the American Chemical Society. 1954b;**76**:6324-6329

[46] Arnon DI. Photosynthetic electron transport: Emergence of a concept, 1949–59. Photosynthesis Research. 1991; **29**:117-131

[47] Arnon DI, Tsujimoto HY, Tang GM. Proton transport in photooxidation of water: A new perspective on photosynthesis. Proceedings of the National Academy of Sciences of the United States of America. 1981;**78**:2942-2946

[48] Arnon DI, Tsujimoto HY, Tang GM-S. Contrasts between oxygenic and anoxygenic photoreduction of ferredoxin: Incompatibilities with prevailing concepts of photosynthetic electron transport. Proceedings of the National Academy of Sciences USA. 1980;**77**:2676-2680

[49] Arnon DI. Divergent pathways of photosynthetic electron transfer: The autonomous oxygenic and anoxygenic photosystems. Photosynthesis Research. 1995;**46**:47-71

[50] Fernandez-Velasco JG, Jamshidi A, Gong XS, Zhou J, Ueng RY. Photosynthetic electron transfer through the cytochrome b6f complex can bypass cytochrome f. The Journal of Biological Chemistry. 2001;**276**:30598-30607

[51] Pesaresi P, Scharfenberg M, Weigel M, Granlund I, Schr€oder WP, Finazzi G, et al. Mutants, overexpressors, and interactors of arabidopsis plastocyanin isoforms: Revised roles of plastocyanin in photosynthetic electron flow and thylakoid redox state. Molecular Plant. 2009;**2**:236-248

[52] Zhang L, Pakrasi HB, Whitmarsh J. Photoautotrophic growth of the cyanobacterium Synechocystis sp. PCC 6803 in the absence of cytochrome c553 and plastocyanin. The Journal of Biological Chemistry. 1994;**269**: 5036-5042

[53] Forsman JA, Fagerlund RD, Biswas S, Summerfield TC, Eaton-Rye JJ. The PsbT protein modifies the bicarbonate-binding environment of Photosystem II. New Zealand Journal of Botany. 2020;**58** (4):406-421. DOI: 10.1080/0028825X.2020.1772320

[54] Emerson R, Chalmers R, Cederstrand C. Some factors influencing the long-wave limit of photosynthesis. Proceedings of the National Academy of Sciences. 1957;**43**:133-143

[55] Emerson R. Dependence of yield of photosynthesis in long wave red on wavelength and intensity of supplementary light. Science. 1957;**125**: 746

[56] Govindjee G, Rabinowitch E. Photosynthesis. 1st ed. New York City: John Wiley & Sons, Inc.; 1969

[57] Manoj KM. Aerobic respiration: Criticism of the proton-centric explanation involving rotary adenosine triphosphate synthesis, chemiosmosis principle, proton pumps and electron transport chain. Biochemistry Insights. 2018;**11**:1178626418818442

[58] Manoj KM, Ramasamy S, Parashar A, Gideon DA, Soman V, Jacob VD, et al. Acute toxicity of cyanide in aerobic respiration: Theoretical and experimental support for murburn explanation. Biomolecular Concepts. 2020a;**11**:32-56

[59] Badger MR, Price GD. The role of carbonic anhydrase in photosynthesis. Annual Review of Plant Physiology and

Plant Molecular Biology. 1994;**45**: 369-392

[60] Kupriyanova E, Pronina N, Los D. Carbonic anhydrase: A universal enzyme of the carbon-based life. Photosynthetica. 2017;**55**:3-19

[61] Moroney JV, Bartlett SG, Samuelsson G. Carbonic anhydrases in plants and algae. Plant, Cell & Environment. 2001;**24**:141-153

[62] Lu YK, Stemler AJ. Extrinsic photosystem II carbonic anhydrase in maize mesophyll chloroplasts. Plant Physiology. 2002;**128**(2):643-649. DOI: 10.1104/pp.010643

[63] Shitov AV, Pobeguts OV, Smolova TN, Allakhverdiev SI, Klimov VV. Manganese-dependent carboanhydrase activity of photosystem II proteins. Biochemistry (Mosc). 2009;**74**(5): 509-517. DOI: 10.1134/s00062979090 50058

[64] Hines KM, Chaudhary V, Edgeworth KN, Owens TG, Hanson MR. Absence of carbonic anhydrase in chloroplasts affects C_3 plant development but not photosynthesis. Proceedings of the National Academy of Sciences. 2021;**118**: e2107425118

[65] Housecroft CE, Sharpe AG. Inorganic Chemistry. 2nd ed. Harlow (England): Prentice-Pearson-Hall; 2005. p. 368

[66] Stumm W, Morgan JJ. Aquatic Chemistry, Chemical Equilibria and Rates in Natural Waters. 3rd ed. New York: John Wiley & Sons, Inc.; 1996

[67] Berg JM, Tymoczko JL, Stryer L. Biochemistry. 5th ed (Section 9.2). New York: W H Freeman; 2002

[68] Gutknecht J, Bisson MA, Tosteson FC. Diffusion of carbon dioxide through lipid bilayer membranes: effects of carbonic anhydrase, bicarbonate, and unstirred layers. Journal of General Physiology. 1977;**69**:779-794

[69] Pocker Y, Bjorkquist D. Stopped-flow studies of carbon dioxide hydration and bicarbonate dehydration in H_2O and D_2O acid–base and metal ion catalysis. Journal of the American Chemical Society. 1977;**99**:6537-6543

[70] Dismukes GC, Klimov VV, Baranov SV, Kozlov YN, DasGupta J, Tyryshkin A. The origin of atmospheric oxygen on Earth: The innovation of oxygenic photosynthesis. Proceedings of the National Academy of Sciences. 2001;**98**: 2170-2175

[71] Bakhmutova-Albert EV, Yao H, Denevan DE, Richardson DE. Kinetics and mechanism of peroxymonocarbonate formation. Inorganic Chemistry. 2010;**49**(24): 11287-11296

[72] Richardson DE, Yao H, Frank KM, Bennett DA. Equilibria, kinetics, and mechanism in the bicarbonate activation of hydrogen peroxide: Oxidation of sulfides by peroxymonocarbonate. Journal of the American Chemical Society. 2000;**122**(8):1729-1739

[73] Patra SG, Mizrahi A, Meyerstein D. The role of carbonate in catalytic oxidations. Accounts of Chemical Research. 2020;**53**(10):2189-2200. DOI: 10.1021/acs.accounts.0c00344

[74] Kuttassery F, Sebastian A, Mathew S, Tachibana H, Inoue H. Promotive effect of bicarbonate ion on one-electron oxidation initiated two-electron water oxidation to form hydrogen peroxide catalyzed by lluminum porphyrins.

ChemSusChem. 2019;**12**(9):1939-1948. DOI: 10.1002/cssc.201900560

[75] Okrasa K, Kazlauskas RJ. Manganese-substituted carbonic anhydrase as a new peroxidase. Chemistry - A European Journal. 2006; **12**:1587-1596

[76] Manoj KM, Gideon DA. Structural foundations for explaining the physiological roles of murzymes embedded in diverse phospholipid membranes. Biochimica et Biophysica Acta – Biomembranes. 2022c;**1864**(10): 183981. DOI: 10.1016/j.bbamem.2022.183981

[77] Manoj KM, Parashar A, Venkatachalam A, Goyal S, Satyalipsu SPG, Gade SK, et al. Atypical profiles and modulations of heme-enzymes catalyzed outcomes by low amounts of diverse additives suggest diffusible radicals' obligatory involvement in such redox reactions. Biochimie. 2016;**125**:91-111

[78] Bielski BHJ, Cabelli DE. Superoxide and hydroxyl radical chemistry in aqueous solution. In: Foote CS et al., editors. Active oxygen in chemistry. New York: Chapman & Hall; 1995. pp. 66-104

[79] Buxton GV, Greenstock CL, Helman WP, Ross AB. Critical Review of rate constants for reactions of hydrated electrons, hydrogen atoms and hydroxyl radicals ($\cdot OH/\cdot O^-$) in aqueous solution. Journal of Physical and Chemical Reference Data. 1988;**17**:513. DOI: 10.1063/1.555805

[80] Jagendorf AT, Uribe E. ATP formation caused by acid-base transition of spinach chloroplasts. Proceedings of the National Academy of Sciences of the United States of America. 1966;**55**(1): 170-177. DOI: 10.1073/pnas.55.1.170

[81] Warburg O. Annual Review of Biochemistry- Prefatory Chapter. Annual Review of Biochemistry. 1964; **33**:1-14

[82] Radmer R, Ollinger O. Isotopic composition of photosynthetic O_2 flash yields in the presence of $H_2{}^{18}O$ and $HC^{18}O_3{}^-$. FEBS Letters. 1980;**110**(1): 57-61

[83] Stemler A, Radmer R. Source of photosynthetic oxygen in bicarbonate-stimulated Hill reaction. Science. 1975; **190**:457-458

[84] Koroidov S, Shevela D, Shutova T, Samuelsson G, Messinger J. Mobile hydrogen carbonate acts as proton acceptor in photosynthetic water oxidation. Proceedings of the National Academy of Sciences. 2014;**111**: 6299-6304

[85] Metzner H. Water decomposition in photosynthesis? A critical reconsideration. Journal of Theoretical Biology. 1975;**51**:201-231

[86] Shevela D, Nöring B, Koroidov S, Shutova T, Samuelsson G, Messinger J. Efficiency of photosynthetic water oxidation at ambient and depleted levels of inorganic carbon. Photosynthesis Research. 2013;**117**:401-412

[87] Van Rensen JJS, Xu C, Govindjee. Role of bicarbonate in photosystem II, the water-plastoquinone oxido-reductase of plant photosynthesis. Physiologia Plantarum. 1999;**105**:585-592

[88] Barber J. Photosystem II: Redox and protein components. Encyclopedia of Biological Chemistry. 2013:512-519

[89] Cohn M, Drysdale GR. A study with O18 of adenosine triphosphate formation in oxidative phosphorylation. The

Journal of Biological Chemistry. 1955; **216**:831-846

[90] Cohn M. A study of oxidative phosphorylation with O-18 labeled inorganic phosphate. The Journal of Biological Chemistry. 1953;**201**:735-750

[91] Manoj KM. Chlorinations catalyzed by chloroperoxidase occur via diffusible intermediate(s) and the reaction components play multiple roles in the overall process. Biochimica et Biophysica Acta/General Subjects. 2006;**1764**: 1325-1339

[92] Manoj KM, Tamagawa H, et al. Murburn model of vision: Precepts and proof of concept. Journal of Cellular Physiology. 2022;**237**(8):333–3355. DOI: 10.1002/jcp.30786

[93] Hussein R, Ibrahim M, Bhowmick A, Simon PS, Chatterjee R, Lassalle L, et al. Structural dynamics in the water and proton channels of photosystem II during the S_2 to S_3 transition. Nature Communications. 2021;**12**(1):6531. DOI: 10.1038/s41467-021-26781-z

[94] Johnson VM, Pakrasi HB. Advances in the understanding of the lifecycle of photosystem II. Microorganisms. 2022; **10**(5):836. DOI: 10.3390/ microorganisms10050836

[95] Jaiswal S, Bansal M, Roy S, Bharati A, Padhi B. Electron flow from water to $NADP^+$ with students acting as molecules in the chain: A Z-scheme drama in a classroom. Photosynthesis Research. 2017;**131**(3):351-359. DOI: 10.1007/s11120-016-0317-z

[96] Mohapatra PK, Singh NR. Teaching the Z-Scheme of electron transport in photosynthesis: a perspective. Photosynthesis Research. 2015;**123**(1): 105-114. DOI: 10.1007/s11120-014-0034-4

Chapter 5

Perspective Chapter: Beyond the Chlorophyll Molecule, Are There Other Organic Compounds Capable of Dissociating the Water Molecule? New and Unexpected Insights

María del Carmen Arias Esparza and Arturo Solis Herrera

Abstract

In the XVII century, researchers throughout Europe began to study the composition of the atmosphere, discerning its physicochemical properties and composition. Since then, it has been observed that the concentration of oxygen in the air around us is relatively low. Lavoisier and Priestley, in the middle of XVII century, observed that plants leaves could replenish oxygen in an impoverished atmosphere. They concluded that chlorophyll possessed the intrinsic property of dissociating the molecule from water. At the XVIII century, the systematic study of human physiology began to deepen, and it was found that the oxygen levels inside the human body were five times higher than those of the atmosphere. The explanation given was that the lung, by means of some unknown mechanism like those of the swim bladder of some fish, was able to concentrate oxygen from the atmosphere and introduce it into the bloodstream. But such a theoretical mechanism has not been found after 200 years of searching. However, there is no way to explain how the concentration of oxygen rises substantially in the tiny distance between the alveolar space and the blood capillaries of the lung. Circumstantially, we found the mechanism during an observational study about the blood vessels entering and leaving the human optic nerve: Our body has several molecules capable of dissociating the molecule from water, such as plants.

Keywords: photosynthesis, oxygen, hydrogen, water, sunlight, hemoglobin, melanin, lignin

1. Introduction

The life cycle is currently interpreted as the concatenation of two processes opposite each other, one is endergonic and the other is exergonic, in one, CO_2 is reduced to glucose; and in the other, glucose is oxidized to CO_2. One happens in the plant kingdom and the other happens in the animal kingdom. The products of the metabolic activity of plants (oxygen and glucose) are harnessed by the animal part (CO_2 and water) and vice versa.

The primary and secondary products of each part of the equation are reused by the other side of the reaction and thus are recycled indefinitely from the beginning of time. Therefore, we are facing a unique sustainability scheme that has allowed life to thrive over eons of years. Thus, photosynthesis and respiration are thus seemingly the reverse of one another [1].

It is postulated that the plant dissociates the molecule of water and obtains oxygen and hydrogen, which is called photosynthesis. Since the notable energy needed to break the water molecule is taken from sunlight because plants have chlorophyll, a molecule whose formula is known quite well, which, using the ends of visible light, collects the enormous energy required to convert liquid water into its gaseous components hydrogen and oxygen. The latter is expelled into the atmosphere for the benefits of living beings because they absorb it from the air that surrounds them through respiration, a process that implies that when expanding our thorax, the air penetrates through the airways, and once inside the lung's functional parenchyma, atmospheric oxygen passively passes through the lung tissues and reaches the blood circulation.

Such passive penetration through lung tissues is called diffusion. It is characterized because it does not require energy to be carried out as it happens spontaneously since the molecules naturally tend to occupy all the space that surrounds them, moving from the area of highest concentration to the one with the lowest concentration (of molecules).

It is important to note that diffusion does not expend energy, because if it were, the already committed theoretical calculations about the energy that our body requires and obtains from the environment to carry out its functions, would be even more confusing. So, let us go in parts. To date, it is not understood how oxygen from the atmosphere passes through the lung tissues and rises inside the body, for example in the blood; until a substantive difference in relation to the proportion of oxygen in the atmosphere is reached.

In the atmosphere, the partial pressure of oxygen is about 21%, but in the blood, saturation reaches more than 95%, and the explanation for this difference has been the subject of debate since the mid-eighteenth century [2].

2. Long-standing theories about oxygen transportation into blood

Transport of O_2 from environment to the mitochondrion is accomplished in an, exhaustively described but not yet well-understood, sequence of diffusive and convective steps [3]. Clinicians and researchers are greatly limited in their understanding of oxygen supply in health and under disease, mainly by technical difficulties to assess organ-specific cellular metabolism perturbations caused by O2 tissues availability, that ideally must be non-invasively methods.

Oxygen is present in alveolar air at about 14% and easily run down to 10 or even 8 percent [4]. There is evidence that the relation between blood supply and ventilation in individual groups of alveoli is not even one. In some alveoli, the oxygen runs down, and CO_2 accumulates faster than others. Hence the oxygen percentage of the mixed alveolar air becomes altogether deceptive as an index of the degree of oxygenation of the mixed arterial blood.

Oppositely, the CO_2 percentage remains a reliable index of the degree of saturation of arterial blood with CO_2 [5]. When the alveolar oxygen pressure reaches about 120 mm Hg, corresponding to about 17% of oxygen in the dry alveolar air, beyond which a further rise in alveolar oxygen pressure has no effect. The oxygen pressure in all the lung alveoli would be at equal high levels at the beginning of apnea but would fall at unequal rates in the different alveoli [6].

Our body seems to ignore oxygen and instead handles CO_2 in a very precise, very careful way, as if it were a life-or-death issue that it is because the high levels of CO_2 in the body quickly impoverish the ability of the cells to dissociate the molecule from the water, with which death happens in less than a minute.

Lung diseases compromise the available respiratory surface to gas exchange of CO_2, which is why they are so serious. Oxygen has no relevance because the cell gets it from the water it contains, like plants.

Interestingly, the alveolar CO_2 percentage goes up as the barometric pressure goes down, that is, during climbing; but the partial pressure of CO_2 remains almost the same in the alveolar air. Alveolar air, with breathing normal, contains about a third less oxygen than the inspired air. It follows that when the oxygen percentage will be reduced to about half, that is, from about 13% of atmosphere to about 6.5% [7]. Such a diminution corresponds to a saturation of about 80% of the hemoglobin with oxygen, and any further diminution will cause a rapid fall in the saturation.

Because that the concentration of oxygen in the air we breathe has a significant effect on CO_2 exchange, our body seems to respond to these fluctuations very precisely, but it is not so much oxygen but CO_2 that the body responds to.

3. Atmospheric oxygen cannot pass lung tissues and reach bloodstream

Oxygen, fortunately, is given by methods that are either ineffective or wasteful. A funnel over the patient's face, a rubber catheter into the patient mouth or nose, an anesthetic mask, etc. In any case, it would be undesirable to continue the administration of pure oxygen for more than a limited time [8].

With 2 liters, a minute the percentage of saturation of the hemoglobin that is, in pneumonia case, rose from 82 to 95%. With 3 liters a minute, the saturation rose to 97%. In the case of exercise, the alveolar CO_2 pressure has a well-marked rise when a little oxygen is added to the inspired air. So, the apparent benefit of oxygen in these cases is that it increases the efficiency of CO_2 expulsion.

So, how do you explain that oxygen rises in the blood by administering supplemental oxygen if since 1850 Ludwig, Bohr [9], Halender [10], and others, found experimentally that oxygen does not pass through the lung tissues, so it is not able to reach the bloodstream?

The answer is oxygen toxicity. For about 400 years, it has been observed that various poisons, such as arsenic, and others, when administered in minimum doses, can induce positive responses in the health of people, for example, elevation of blood oxygen levels. Therefore, when we administer supplemental oxygen, which is toxic and not absorbed, the tissues are harassed, and up to a certain limit, the tissues respond by increasing the rate of turnover of water dissociation. So, the increase in % SpO_2 that follows the administration of supplemental oxygen is oxygen that comes from the water inside the cells, not from the atmosphere [11].

Clinical observations in cases of CO poisoning, such as although the amount of oxygen transported by the blood is diminished, the oxygen pressure in the arterial blood remains normal. It is something difficult to explain, but if we start from the fact that hemoglobin does not transport oxygen or CO, nor CO_2, but hemoglobin is one of the molecules that the human body has that are able to dissociate the water molecule, then it is explained in a congruent, coherent way, without complicated mathematical models, such as Krogh's [12], which to top it off is totally theoretical, is completely imaginary, and handles theoretical concepts so far-fetched in its eagerness to explain the supposed transport of oxygen through the lungs, that it cannot even be contrasted experimentally. Considerable discrepancy was found between pO_2 microsensor data and results from that model (**Figure 1**) [13].

The cylindrical steady-state model developed by Krogh with Erlang has served as the basis of understanding oxygen supply in living tissue for over eighty years. It has been extensively used for situations such as drug diffusion, water transport, and ice formation in tissues. However, the applicability of the model to make even a qualitative prediction of the oxygen level of specific volumes of the tissue is still controversial, which is paradoxical because supposedly it was developed. And after more than 100 years and a myriad of equations that have been gradually added with the intention of making it functional or at least believable, the controversy continues about the fact that atmospheric oxygen passes through the lungs and reaches the bloodstream by simple diffusion (**Figures 2** and **3**).

The reason is that the mathematical model of Krogh-Erlang tries to explain something that is not physic-chemically possible, thereby is not real, but it is a very rooted dogma that cannot be even experimentally contrasted [15]. The mathematical and statistical models that are used to try to explain biological processes, such as gas exchange, usually do not work because in biology the variables are continuous random (nonlinear behavior) in nature.

If we applied the model of Krogh-Erlang to study the effect of acclimatization to high altitude, we have that at high altitude at rest, arterial carbon dioxide tension, oxygen saturation, and oxygen tension were significantly reduced due to several factors, for instance, the amount of light is diminished in extreme weather (cold) the capacity of human body to take oxygen from intracell location is impaired; and oppositely, the oxygen content is increased, given that hemoglobin concentration increase, recall hemoglobin dissociates the water molecule unreversible [16].

Notice the explanation: arterial oxygen content was increased because of an increase in the hemoglobin concentration, which reductionistic alludes to the fact that

$$[O_2] = \alpha P \quad Hb+nO2\rightleftarrows kk'Hb(O2)n \quad S=K[O2]n1+K[O2]n \quad S=(P/P50)n1+(P/P50)n$$

$$O2+Hb4(O2)i-1\rightleftarrows kiki'Hb4(O2)i,i=1,2,3,4 \quad S=(\sum i=14i[Hb4(O2)i])/(\sum i=14[Hb4(O2)i])$$

$$S=a1P+2a2P2+3a3P3+4a4P44(1+a1P+a2P2+a3P3+a4P4) \quad a_l = \alpha^l K_1 K_2 \ldots K_l$$

$$S = (S_m - S_o)\exp[-(R/K)\exp(-KP)] + S_o \quad P = \ln(R/K) - \ln(\ln[(S_m - S_o)/(S - S_o)])/K$$

$$O2+Hb\rightleftarrows kk'HbO2 \quad R = k'[O_2][Hb] - k[HbO_2] \quad ddt[O2]=-R,ddt[Hb]=-R,ddt[HbO2]=R$$

$$S=K[O2]1+K[O2] \quad k'=k1\alpha P(PP50)n \quad R=k[HbT](F(P)1-F(P)(1-S)-S)$$

$$Ri=ki'[O2][Hb4(O2)i-1]-ki[Hb4(O2)i],i=1,2,3,4 \quad Mb+O2\rightleftarrows kk'MbO2$$

$$S=[MbO2][Mb]+[MbO2]=K[O2]1+K[O2] \quad j=-D\partial c\partial n \quad j = -D\nabla c \quad \partial c\partial t=\nabla D\nabla c+R$$

$$j = D_o\nabla[O_2] + D_{Hb}\nabla[HbO_2] \quad j = D_o\nabla[O_2] + D_{Hb}[Hb_T]\nabla S \quad j=(Do+DHb[HbT]dSd[O2])\nabla[O2]$$

$$1+DHb\alpha Do[HbT]dSdP \quad \nabla[HbO2]=[HbT]\alpha-1dSdP\nabla P \quad \gamma=(LL\beta)2 \quad c = 1 \text{ at } r = R_L$$

$$Pe(u\partial c\partial x+v\partial c\partial r)=\partial 2c\partial r2+1r\partial c\partial r+\partial 2c\partial x2 \quad c = 0 \text{ at } x = -1 \text{ and } x = 1$$

Figure 1.
A sample of the first 32 equations of already 120 described that have been implemented with the aim of building Krogh's acceptable theoretical (imaginary) model of oxygen transportation theory [14].

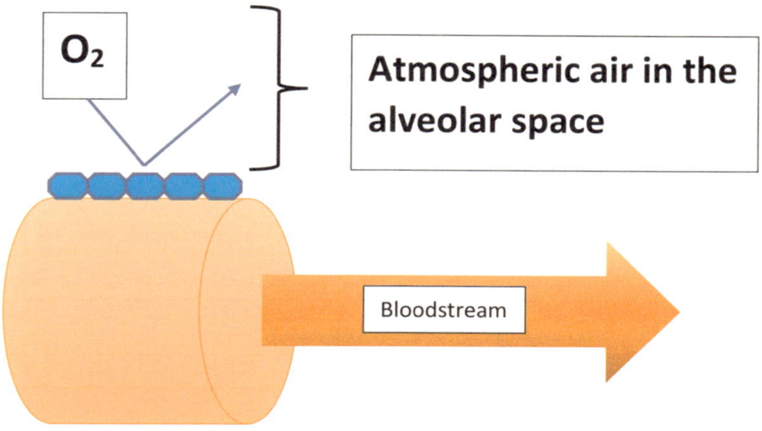

Figure 2.
Represents pulmonary alveolar cells, with about 70% water, thereby oxygen from alveoli air is repelled and cannot reach the bloodstream. Represents blood vessels.

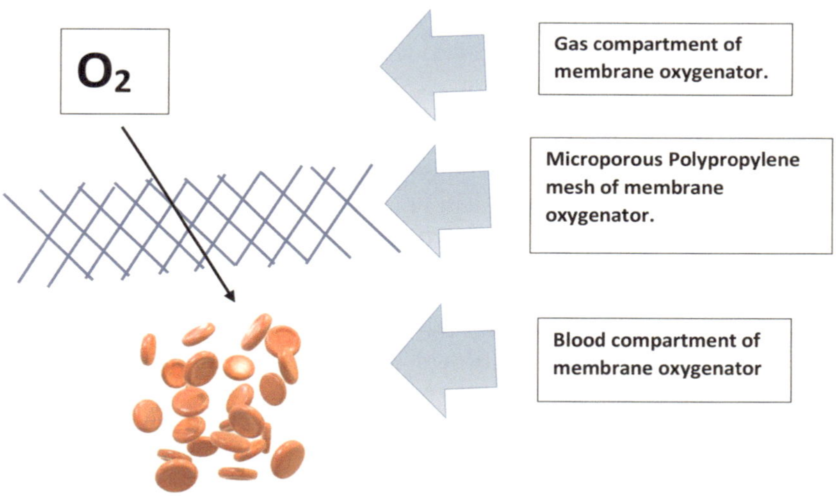

Figure 3.
In membrane oxygenators used for extracorporeal circulation, the story is different, because when the barrier that represents the intracellular water disappears, the oxygen easily crosses the microporous polypropylene mesh reaching the blood compartment.

there is more hemoglobin there is more oxygen transport from the atmosphere to the tissues, despite since the second half of the nineteenth century it was reported that it was not possible. So, if we interpret the results from our point of view which is that hemoglobin molecules possess the unexpected ability to dissociate the molecule from water, then the increase in hemoglobin concentration explains the increase in the body's ability to dissociate the molecule and therefore maintains blood oxygen saturation levels properly.

With the previous paragraphs we intend to demonstrate, although briefly, that oxygen from the atmosphere is not absorbed as it has been thought for at least three centuries. The high levels of oxygen normally present inside the body come from the water that the cells contain inside and not from the atmosphere that surrounds us. Therefore, we can discard the ancestral belief that plants expel oxygen into the atmosphere for the benefit of other living beings, because plants expel oxygen because they also take it from the water they contain inside, they use what is necessary for their metabolic processes and expel the surplus, given the significant toxicity of oxygen.

Once we are aware that we do not take oxygen from the atmosphere that surrounds us, but from the water that each cell contains inside, then we can move on to the following:

4. Glucose is not a source of energy

Following the scheme of the textbooks, it was thought that once the oxygen from the atmosphere was absorbed by the lungs and reached the bloodstream, then it was distributed to all the cells of the organism so that, when combined with glucose, in a kind of strange, graduated combustion, by which the body obtains energy in the form of heat, which is the most inefficient form of energy.

But just as oxygen transport was full of postulates and imaginary equations, we must do the same thing when trying to explain how our organism obtains energy from combining glucose with oxygen. To begin with, it is assumed that light energy is stored in the covalent bonds of glucose, which is indefensible, given that energy cannot be stored.

But if we ignore that small detail, then what follows, which is also another assumption about the eukaryotic cell, through an organelle called mitochondria, can recover the energy stored in the covalent bonds of glucose. Of course, it is also through mechanisms that are not understood, they are only imaginary.

Let us remember that the study of metabolism has a long history, the roots of which can be traced back to the year 1614 when Santorio Sanctorius published his results on body weight fluctuations during the course of a day [17]. Until today, the era of pathway databases are going through the realization in the nineteenth century that the reactions within a cell are the same as those studied in chemistry, the discovery that enzymes catalyze metabolic reactions, the first complete metabolic pathways (1930), among which the tricarboxylic acid (TCA) cycle or Krebs cycle (1937) [18].

5. The Krebs cycle is present in every cell that uses oxygen to produce energy

From 1930 onwards, an increasing number of metabolic pathways have been unraveled but misunderstood because often studied in isolation [19]. Pathways interact in a highly complex way and together constitute the metabolic network. This highly organized and complex network of reactions, astonishing can adapt to a constantly changing environment, through mechanism poorly understood, thereby, it is still a mystery.

Although the application of mathematical models to the study of biological phenomena has demonstrated time and again their futility, one of the ultimate goals is to construct the full metabolic network in all its detail so can be implemented a complete or almost complete mathematical model that can be used to generate experimentally verifiable hypotheses, perhaps identify potential drug targets, and ideally but very difficult to reach, to simulate the effect of network perturbations, such a loss of function [20].

But one of the problems faced by the promoters of the so-called metabolic pathway databases complexity of human metabolism is how to represent knowledge without affecting the ability of a pathway database to capture the biological complexity of human metabolism. For instance, which aspect of the metabolic network is important and to what detail it needs to be represented, which is furthermore complicated by the changing nomenclature of enzymes and metabolites in the course of time [21].

And worst, not every piece of the metabolic network, conclusive evidence is available, and some parts might still be subject to controversy. It is not easy to increase the awareness of the scientific community of the existing differences and biological inaccuracies within the descriptions provided by pathway databases, and the best way to resolve them.

One example is the description of TCA cycle as given by ten databases. None of these were entirely consistent with the literature, and the biochemistry behind the TCA cycle turned out to be not as clear-cut as one might expect. The various human metabolic networks described by these databases have not been systematically compared, nor has the extent to which they differ been quantified. So, it is no surprise that the degree of consensus among them is disappointingly low.

Reactions are not always balanced, especially with respect to electrons (e^-), protons (H^+), and water (H_2O) [22]. So, when we analyzed the Krebs cycle, already described in 1937, the consensus on reaction level is surprisingly low. The databases agree on 5 of the 30 reactions that mold up the Krebs cycle.

Therefore, it is not surprising the significant differences between the different biochemistry textbooks, since each author explains in his own way a theory that tries to explain how the cell obtains energy from its environment. But, if glucose were a source of energy, diabetic patients would be able to fly.

So, since it is not true that our body takes oxygen from the air, then it is also not true that our body takes energy from glucose. Our discovery that there are several molecules in our body capable of dissociating water [23], such as plants, also explains the hitherto mysterious origin of energy in eukaryotic cells.

6. When the water molecule is dissociated, energy is released

When energy is applied to a molecule, sooner or later, it breaks down. And in the case of water, the same thing happens. The dissociation of the water molecule is highly endergonic, that is, it requires a lot of energy. If we want to dissociate the water in the laboratory, we must heat the water to two thousand degrees celsius.

But nature carries it out in a very precise way and at room temperature through various molecules. The best known to date is chlorophyll, which dissociates the water molecule irreversibly, which is represented as follows:

$$H_2O_{(liq)} \rightarrow 2H_{2(gas)} + O_{2(gas)} \tag{1}$$

So, when our body dissociates the molecule from water in the exact way it has done since the beginning of time, it gets two fundamental elements, such as oxygen and hydrogen. And part of the energy that is released when breaking the water molecule is captured and transported by hydrogen, remember that this element is the main carrier of energy in the entire universe, so our body cannot be different.

So, by dissociating water, our body gets the two basic elements in the biochemical sequence of life. That is why the dissociation of water is considered the very first reaction of life in plants, and from now on, also in humans or mammals.

Of the several molecules that we have found in the human body that can dissociate irreversible the water molecule, we have hemoglobin, myoglobin, cytochrome P 450, and bilirubin; but the main one for us is melanin because until now it is the only one that dissociates and re-associates the water molecule using as a source of energy the sunlight. The reaction would be described as follows:

$$H_2O_{(liq)} \rightarrow 2H_{2(gas)} + O_{2(gas)} \rightarrow H_2O_{(liq)} + 4e^- \quad (2)$$

For every two water molecules that are re-formed, 4 high-energy electrons ($4e^-$) are generated. The turnover rate of the process inside melanin is in the range of nano and picoseconds. The presence of high-energy electrons can be verified in relatively easy ways (**Figures 4** and **5**).

On the other hand, hydrogen generation from melanin can also be easily demonstrated (**Figures 6–8**).

Melanin molecule is the perfect candidate to be the much sought-after dark matter of the universe, as it meets the three main requirements: 1) extraordinary stability, 2) massive nature, and 3) absorption capacity of any form of energy (**Figures 9** and **10**).

1. Extraordinary stability - melanin is the most stable molecule known, proven 160 million years [24].

2. Both the formula and thereby molecular weight of melanin is unknown; however, the molecular weight is estimated in millions of Daltons [25].

3. Melanin is the darkest substance man has ever known [26] because it absorbs any type of energy and dissipates it in a unique way: dissociating the water molecule [27].

The first part of this work was about breathing with the intention of making the reader aware that our body does not take oxygen from the air around it but from the

Figure 4.
LED lights are energized with a melanin-based battery.

Figure 5.
Depending on the geometry of the melanin-based battery, we obtain direct current values such as the one shown in the photograph.

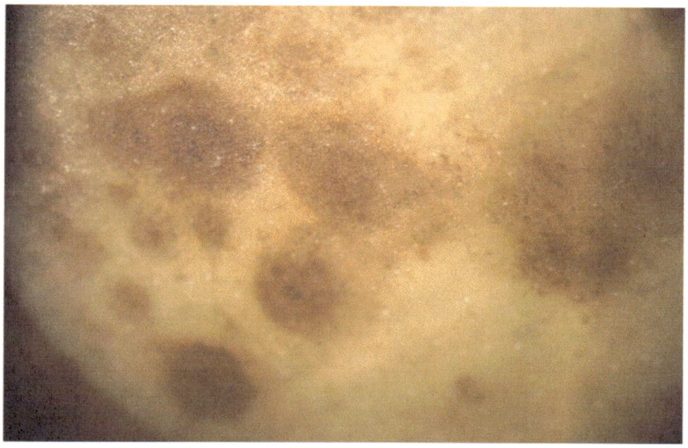

Figure 6.
Melanin of the skin, observed through light biomicroscope (5X).

water that our cells contain. A mistaken belief that has finally significantly disrupted the development of knowledge about mammalian biology.

Since the late nineteenth century, Christian Bohr, based on his own experimental studies, concluded that the oxygen we have inside the body did not come from the atmosphere by simple diffusion, but rather seemed a secretion of lung tissues [28].

This means that since then Christian Bohr inferred that we do not take oxygen from the atmosphere around us. He could not go further because apparently, at that time, there was no way to explain the origin of this oxygen. Fortunately, now we have the explanation about oxygen secretion in the lungs and all cells, which also explains the functioning of the swim bladder in fish.

7. Photosynthesis in plants

Currently, the most studied non-trivial quantum effects in biological systems are photosynthesis and the magnetic orientation of birds [29]. Quantum effects are often

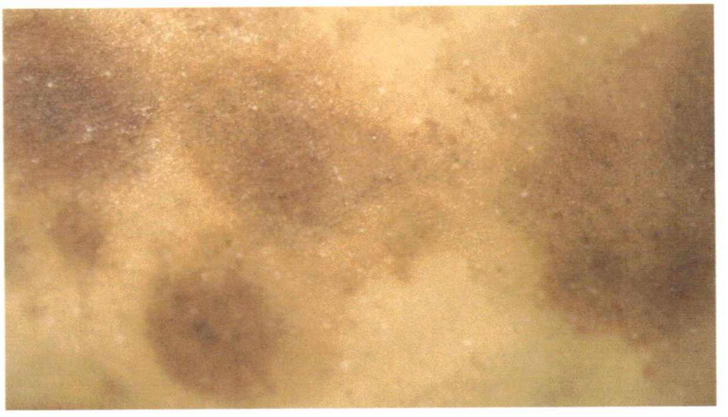

Figure 7.
The melanin of the skin observed through a light biomicroscope with greater magnification (10X).

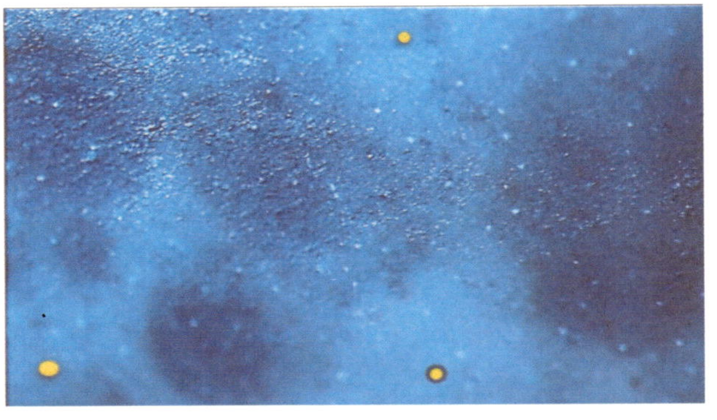

Figure 8.
By illuminating melanin with UV light (10–300 nm) hydrogen fluoresces (yellow dots). Oxygen does not fluoresce (10X).

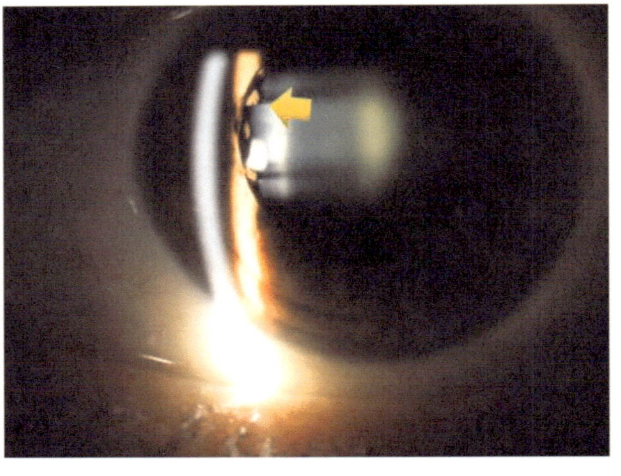

Figure 9.
In some patients, the pigmented cells (with melanin) tend to form sprout-like formations in the iris edge (yellow arrow).

(NASA ESA CSA and STScI)

Figure 10.
The color of melanin hardly changes whether it is a few cm from us, as in ***Figure 8****, or thousands of light-years away, for example, in stellar space.*

associated with paradoxes (Schrodinger's cat, Wigner's friend), but paradoxes are also present in the understanding of living systems (e.g., Levinthal's paradox). The question arises whether the solutions to paradoxes (unsolved problems) in biology may relate to quantum mechanics.

The non-trivial role of quantum mechanics in living systems can be applied to the subject of studies for those components of living systems in which these effects play an important role. These effects include photosynthesis, protein folding, molecular recognition, and others.

Every organism is a complex network of chemical reactions, the effective functioning of which is largely due to the transfer of electrons from one atom to another. In this sense, quantum mechanics always plays an important role in energy transmission and conversion in a cell [30]. Non-trivial quantum effects were not known for a long time. The most famous quantum effects are associated with quantum energy conversion in photosynthesis and in the magnetic orientation of birds. On the other hand, the efficiency of energy conversion and transport of substances in the cell is high, and this fact requires explanation, and the involvement of quantum models is necessary to explain this phenomenon.

Quantum mechanics in general is nonlinear. The problem of measurement in quantum mechanics. is the paradox of the collapse of the wave function. The linear Schrodinger equation is only an approximation, and such processes are described in which the fields are classical [31]. The linearity of quantum mechanics is the main cause of the paradox of the collapse of the wave function, taking account of nonlinearity eliminates the paradox.

Quantum coherence was found in the molecules that are directly involved in the process of photosynthesis. A protein antenna receives energy, and this exciton energy is transferred with high efficiency to the specific pigmented protein complexes. The essence of these effects is that a photosynthetic bacteria complex at 77 K showed properties of quantum entanglement under laser excitation. The authors concluded that the quantum effects could be observed at room temperature. These effects may increase the efficiency of energy conversion in photosynthesis [32].

The photosynthetic reaction center is considered a quantum heat engine. It is noted that the equations describing the operation of such machines are similar in many

Figure 11.
Melanin is placed inside the trunk; it is called lignin. However, its function is the transduction of radiant energy to chemical energy through the dissociation and re-form to the water molecules. Thereby, this chemical energy transported by molecular hydrogen impels the growth of plant trunk.

respects to the equations for a laser. Coherence violated detailed balance, thereby increasing the efficiency of the process. A photosynthetic complex (quantum heat engine) converts high-energy photons to a low-entropy flow of electrons [33].

Disposition of melanin or any other photopigment in nature, inside of a cylindrical structure is quite frequently probably explained by the effect of cylindrical lenses that focused the light on form of line (**Figure 11**).

The union of two cylindrical flat lenses concentrates the luminous energy along a line that illuminates (energizes) the total structure, we can observe it from tiny structures such as mitochondria (A), the long bones that in this way illuminates (energize) the entire bone marrow they contain (B), and the blood vessels that when joining two flat-cylindrical lenses, they illuminate the entire path of the blood vessel, energizing the erythrocytes and other blood components (C). Something similar happens with the trunk of trees and the stem of plants because the area of greatest energy is precisely the center of the structure along its entire extension (**Figure 12**).

8. The problem of protein folding

Protein folding is one of the most important problems in molecular biology. This problem (Levinthal's paradox) has been discussed repeatedly [34]. However, despite a large number of publications on this issue, there is no agreement among researchers not only as to what the solution to the paradox is but even whether the solution exists. The first estimates by Levinthal (1968) [35], found that the average time of folding of a long protein molecule should be exponentially large because of the large number of conformational degrees of freedom. Levinthal concluded that a random search is not realized. What then is the mechanism of folding?

The total number of conformational states of a protein chain can be estimated according to (Berezovsky and Trifonov, 2002) [36]: 3^{N1} where N_1 is the number of protein domains. Here it is assumed that each protein domain has three different conformations. If we take the largest possible population of macromolecules (proteins) to be 10^{50} (the mass of these molecules is greater than the mass of earth).

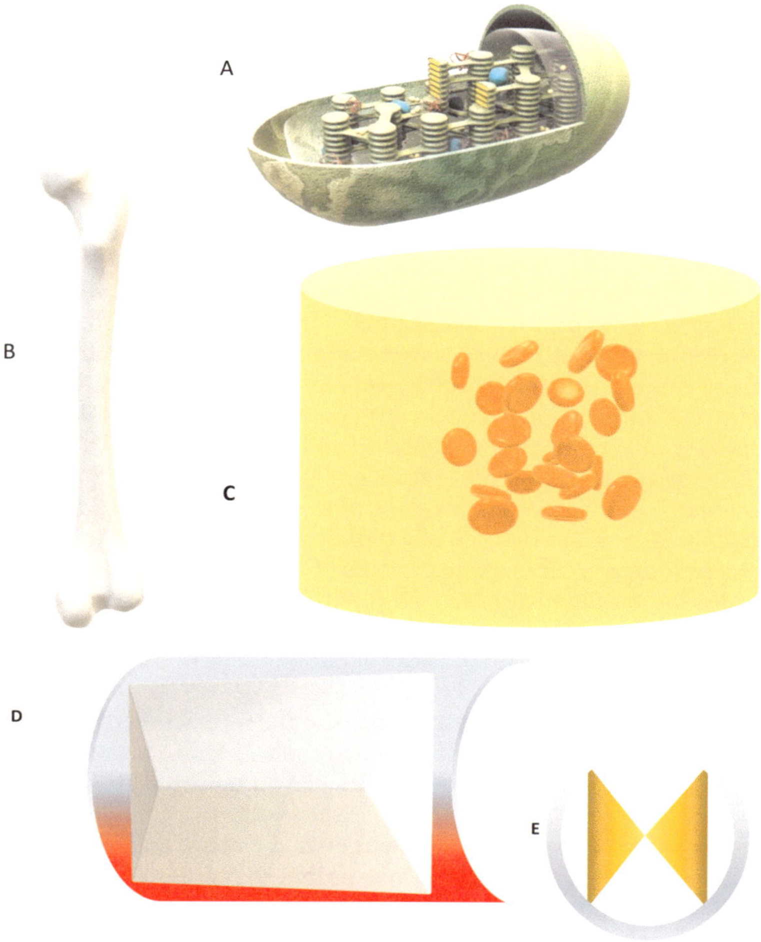

Figure 12.
The cylindrical shape can be interpreted as two flat-cylindrical lenses, which instead of focusing light on one point (like spherical lenses), form a continuous line along the entire cylinder (D, E).

One possible solution to the Levinthal paradox is the assumption of the existence of a funnel-like landscape of the free energy of the protein. This smooth landscape leads to the fact that the characteristic time of protein folding becomes small. Hamiltonian of the interaction between proteins domains contains mainly hydrogen bonding and hydrophobic interaction forces between atoms domains [37]. This model implicitly assumes the existence of a purpose of folding. The author argues that evolution has no goals, and therefore there are no right or wrong results [38].

Transport of molecules and ions is accompanied by processes like protein folding [39].

Recognition of molecules at their reaction sites inside a cell is essential for a cell's normal functioning. It is assumed that reactions between biologically important molecules obey the principle of "lock and key" (or "hand glove"). Otherwise, the stable and precise work of a cell would be simply impossible because of the large number of "wrong" reactions. During the reaction between the key and lock a molecular complex forms. We show that the problems arising in the formation of such a complex are directly related to the problem of protein folding. The formation of complexes may be useful as well as harmful to cells. In the first case, these complexes can serve to transmit energy (information) within the cells on the key-lock principle. In the second case, erroneously emerged complexes, on the contrary, will impede the normal functioning of cells.

It was shown that in the absence of special restrictions, the transport of proteins (and RNA) will be unstable with respect to their interactions with other proteins, etc. Even though most of the positive and negative ions in the cell are quite simple (e.g., sodium, potassium, chlorine), their transportation can also be unstable. Transport systems of any ions are proteins of sufficiently large mass. In this case, the problem remains: why does the transport protein with overwhelming probability transfer the ion from one side of a membrane to another (with the assistance, for example, of ATP) and is not involved in any other reaction? [40].

Experiments show that the effectiveness of the active transport of ions is high and is approximately 90%. The existence of a plurality of spatial structures of biologically important molecules, as well as many variants of chemical reactions between them, is one of the most important obstacles to the functioning of the cell. There must be some special mechanism that significantly restricts the choice of variants in such a system. Without this mechanism, we cannot speak about any significant effect of molecular machines.

One of the most important quantum effects in the interaction of biologically important molecules is the tunnel effect, which can accelerate reactions by orders of magnitude [41]. Thereby, quantum mechanics plays a nontrivial role in various life processes.

Electromagnetic potential may be present even if electric and magnetic fields are absent, but the converse is not true. This means that the system can affect the status of other systems through the electromagnetic potential. That is, such a system "feels" the environment through the electromagnetic potent.

The process of the evolution of life largely depends on the properties of atoms and molecules, which, in turn, are determined by the laws of quantum mechanics [42]. The probability of the evolution of a complex system that can reproduce is exponentially small [43].

9. There are substantial coincidences between plant and animal cells

As is known, the genetic code of all living beings is the same (although there are slight differences in the codes of mitochondria and chloroplasts). However, the mechanism of formation of a single genetic code is still not clear. Initially, it was thought that the genetic code is a "frozen accident"; in recent decades, there are more arguments that the code is the result of evolutionary optimization [44]. Quantum properties that accompany the transport and conversion of energy in electron transfer could occur in the earliest stages of evolution [45].

It should be noted that the creation of copies for a quantum system obeys certain restrictions. One of the basic properties of quantum information is that an unknown state may not be copied, that is, to produce an exact copy we must have complete information about the system [46].

DOI: http://dx.doi.org/10.5772/intechopen.108545

It is not yet determined how important is quantum mechanics to biochemistry. It is possible that prebiotic reactions use quantum effects and then become part of the machinery of DNA [47]. Indeed, a better understanding of quantum mechanics will contribute to a better understanding of the origin of life. Quantum entanglement of protons and other components must be viewed in a broader context [48].

The mechanism of evolution requires the use of a priori information explicitly; otherwise, selection on the molecular level cannot be realized. The problem of measurement in quantum mechanics is still unsolved, and the role of the observer is still unclear [49]. Biology is fundamentally contextual: bio-systems adapt permanently their behavior to contexts.

There are significant similarities between plant and animal cells. Plant and animal cells are both types of eukaryotic cells, meaning they both contain a true nucleus as well as other membrane-bound organelles. The similarities between both are enough to make it believe that in both, the interchange and flow of energy with the surroundings is quite identical, that is, glucose is the source of energy. Undoubtedly, all cells need energy to grow and function, and animal and plant cells both obtain this energy from cellular respiration, a theoretical model. This imaginary metabolic process supposedly takes place in the mitochondria of plant and animal cells and involves the breakdown of glucose to release energy.

And although the chloroplast has been studied exhaustively, since it is considered the only place where the molecule of water dissociates irreversibly, to date, the mechanism by which water is separated into its gaseous components is not yet understood. A lot of emphases has been placed on the oxygen that cells get by dissociating water, which of course has very exact and complex functions, but the cell does not use it to combine it with the glucose because it would be a waste.

Even more so, if the cell gets hydrogen at the same time as oxygen. The energy that is released by breaking the water molecule is captured in some proportion by hydrogen to be transported throughout the interior of the cell. Gases do not combine with water, so both oxygen and hydrogen move through the cytoplasm of the cell, passing through it from where the water is dissociated, usually near the cell nucleus, and following the laws of simple diffusion, it goes to the areas of lower concentration, which in this case would be the periphery, that is, the cell membrane, which also requires the energy that hydrogen carries.

In their displacement from the areas of greatest concentration, that is, where they are produced (chloroplasts in plants, melanosomes in mammals) to the areas of lower concentration (the periphery of cells), the different molecules and intracellular components are capturing either oxygen or hydrogen, with their valuable energy load. What constitutes a very efficient system of both production and distribution of energy.

Remember that the dissociation of water is a process that requires a lot of energy (endergonic), that is two thousand degrees celsius in the laboratory. But this energy, for living beings, is free because they get it from sunlight. So, the very origin of life is blooped in three components: a continuous energy source such as the sun, a unique substrate such as water, and finally the presence of molecules to transduce light energy into chemical energy by dissociating water. Thereby, glucose can be considered the universal precursor of any organic molecule in both plants and animals, but it is not able to provide the energy that its own metabolism requires.

It is time to shake off dogmas that have prevailed for a century. The body does not use a process as crude as combustion or combination with oxygen to obtain energy, let alone heat, then the most inefficient form of energy.

Nature is much more subtle because it takes advantage of a form of free energy, such as sunlight, and transforms it into another form of energy that can be used by different cells than living entities. And the dissociation of the water molecule is by far, the main mechanism, we could say is the universal mechanism. **Figures 11** and **12**.

The molecular formula for Chlorophyll is $C_{55}H_{72}MgN_4O_5$ which astonishingly sums to 137 atoms (55 + 72 + 1 + 4 + 5 = 137). The chlorophyll molecule consists of a central magnesium atom surrounded by a nitrogen-containing structure called a porphyrin ring; attached to the ring is a long carbon-hydrogen side chain, known as a phytol chain. Chlorophyll is the major pigment used by plants for capturing light energy. A chlorophyll molecule consists of a porphyrin head (four pyrrole rings containing nitrogen arranged in a ring around a magnesium ion) and a long hydrocarbon tail.

Chlorophyll is related to human blood: Most chlorophylls are classified as chlorins, which are reduced relatives of porphyrins (found in hemoglobin or blood). This means that under a microscope, the mandala of the red human blood and green chlorophyll are identical except for their central atom. This is an incredible link we have to Plants.

Finally, to the various similarities between plants and animals, we must include the fact that we do not take oxygen from the air around us, but from the water contained in the cells that make us up, such as plants. This implies rewriting biochemistry, biology, physiology, medicine, etc., but we cannot ignore such a transcendent observation.

10. Conclusion

Plants, like mammals, do not take oxygen from the surrounding air, instead, they take the oxygen from water inside the plant. Thereby, the oxygen that leaves expel into the air, coming from intracellular water. The basic mechanism is like animal kingdom: dissociating the water molecule.

Chlorophyll is the utmost studied part of plants, believing that in that place the water dissociation takes place. However, in the trunk, occurs a more efficient form of

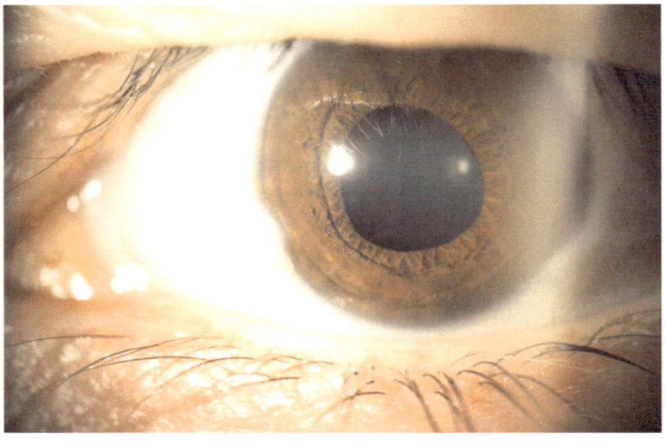

Figure 13.
The human eye contains 40% more melanin than the skin. The photograph shows the anterior surface of the iris, whose resemblance to the surface of some seeds, which also contain melanin, is striking.

Figure 14.
Peach seeds bear a huge resemblance to the anterior surface of the iris, both in shape and with melanin content. And the function of melanin is the same, transforming sunlight into chemical energy through the dissociation of water, which drives each one of the reactions that finally generate the hatching of the seed.

dissociation of water, this is, the dissociation and re-form of the water molecule, while in the leaves, water dissociation is irreversible, like in hemoglobin molecule.

Lignin is a kind of melanin placed in the trunk (**Figure 11**), and like melanin in mammals, has the unsuspected capacity to dissociate and re-form the water molecule. And this unsuspected source of energy and oxygen will allow implementing a better scheme for energy flow in the plants. The current metabolic schemes are biased in trying to frame the dogmas, for example, when the ATP molecule is formed, energy is released as in any molecule, and when the ATP molecule is hydrolyzed to ADP. energy is absorbed.

And yet, in biochemistry textbooks, it is described in the opposite way, which is contrary to reality.

Haldane, in 1913, postulated the controversial notion of the lugs as an oxygen-secreting organ. In that same year appeared mechanism, lie, and personality, in which Haldane declares that "The phenomena of life are of such a nature that no physical or chemical explanation of them is remotely possible" (**Figures 13** and **14**) [50].

Acknowledgements

This work was supported by an unrestricted grant from Human Photosynthesis™ Research Centre. Aguascalientes 20000, México. comagua2000@yahoo.com

Author details

María del Carmen Arias Esparza* and Arturo Solis Herrera
Human Photosynthesis™ Research Centre, Aguascalientes, México

*Address all correspondence to: comagua2000@yahoo.com; carmenrpp@yahoo.com

IntechOpen

© 2022 The Author(s). Licensee IntechOpen. This chapter is distributed under the terms of the Creative Commons Attribution License (http://creativecommons.org/licenses/by/3.0), which permits unrestricted use, distribution, and reproduction in any medium, provided the original work is properly cited.

References

[1] Johnson PM. Photosynthesis. Essays in Biochemistry. 2016;**60**:255-273. DOI: 10.1042/EBC2016001 Portland Press

[2] Gjedde A. Diffusive insights: On the disagreement of Christian Bohr and august Krogh at the centennial of the seven little devils. Advances in Physiology Education. 2010; **34**(4):174-185. DOI: 10.1152/advan.00092.2010

[3] Wagner PD. Limitations of oxygen transport to the cell. Intensive Care Medicine. 1995;**21**(5):391-398. DOI: 10.1007/BF01707407

[4] Douglas CG, Haldane JS. The regulation of normal breathing. The Journal of Physiology. 1909;**38**(5): 420-440. DOI: 10.1113/jphysiol.1909.sp001315

[5] Meakins J, Davies HW. Observations on the gases in human arterial and venous blood. The Journal of Pathology. 1920;**23**(4):451-461. DOI: 10.1002/path.1700230410

[6] Douglas, Haldane. The causes of absorption of oxygen by the lungs. Journal of Physiology. 1999;**XXXVIII**: 401

[7] Mosso A, Kiesow EL. Life of Man on the High Alps. London: T.F. Unwin Edition; 1898

[8] Haldane JS. Respiration. Yale University Press; 1922. p. 172

[9] Bohr C, Nagel WA. Handbuch der Physiologie des Menschen. In German / Deutsch Handbuch der Physiologie des Menschen. Vieweg Edition; 1905

[10] Haldane J, Smith JL. The Absorption of Oxygen by the Lungs. Belfast: Pathological Laboratory, Queen's College; 1898

[11] Herrera AS, Del C A Esparza M, Md Ashraf G, Zamyatnin AA, Aliev G. Beyond mitochondria, what would be the energy source of the cell? Cent Nerv Syst agents. Medicinal Chemistry. 2015; **15**(1):32-41. DOI: 10.2174/1871524915666150203093656

[12] Schubert RW, Zhang X. The equivalent Krogh cylinder and axial oxygen transport. Advances in Experimental Medicine and Biology. 1997;**411**:191-202. DOI: 10.1007/978-1-4615-5865-1_23

[13] Grinberg O, Novozhilov B, Grinberg S, Friedman B, Swartz HM. Axial oxygen diffusion in the Krogh model: Modifications to account for myocardial oxygen tension in isolated perfused rat hearts measured by EPR oximetry. Advances in Experimental Medicine and Biology. 2005;**566**:127-134. DOI: 10.1007/0-387-26206-7_18

[14] Solís Herrera A, Arias-Esparza MC. Oxygen metabolism as a modifiable factor for dementia improvement. In: Chapter in the Book: Neuropsychology of Dementia. London, UK, London, UK, In press, to be published at Aug 05: Intechopen; 2022

[15] Popel AS. Theory of oxygen transport to tissue. Critical Reviews in Biomedical Engineering. 1989;**17**(3):257-321

[16] Moller K, Paulson OB, Hornbein TF, Paulson AS, et al. Unchanged cerebral blood flow and oxidative metabolism after acclimatization to high altitude. Journal of Cerebral Blood Flow & Metabolism. 2002;**22**:118-126

[17] Sanctorius S. Ars de statica medicina. 1614

[18] Krebs HA, Johnson WA. The role of citric acid in intermediate metabolism in animal tissues. Enzymologia. 1937;**4**: 148-156

[19] Alabduladhem TO, Bordoni B. Physiology, Krebs cycle. In: StatPearls [Internet]. Treasure Island (FL): StatPearls Publishing; 2022 [Updated 2021 Nov 21]. Available from: https://www.ncbi.nlm.nih.gov/books/NBK556032/

[20] Oberhardt MA, Palsson BØ, Papin JA. Applications of genome-scale metabolic reconstructions. Molecular Systems Biology. 2009;**5**:320

[21] Stobbe MD. The Road to Knowledge: From Biology to Databases and Back Again. Amsterdam, The Netherlands: University of Amsterdam, UvA-DARE (Digital Academic Repository); 2012. Retrieved from: https://hdl.handle.net/11245/1.385827. May 16 2022

[22] Ott M, Vriend G. Correcting ligands, metabolites, and pathways. BMC Bioinformatics. 2006;**7**:517

[23] Herrera AS. The unsuspected intrinsic property of melanin to transform light into chemical energy and the seed growth. In: Rigobelo EC, editor. Plant Growth [Internet]. London: IntechOpen; 2016 [cited 2022 Jun 06]. Available from: https://www.intechopen.com/chapters/51547. DOI: 10.5772/64542

[24] Colleary, Caitlin. Dolocan Andrei, Gardner James, Singh Suresh, et al. Chemical, experimental, and morphological evidence for diagenetically altered melanin in excepcionally preserved fossils. 12592-12597 | PNAS | 2015 vol. 112 no. 41. DOI: www.pnas.org/cgi/doi/10.1073/pnas.15098311122

[25] Lea AJ. Molecular weight of melanin. Nature. 1947;**159**(4051):843. DOI: 10.1038/159843a0

[26] Hamanaka T, Nishizawa K, Sakasegawa Y, Oguma A, Teruya K, Kurahashi H, et al. Melanin or a melanin-like substance interacts with the N-terminal portion of prion protein and inhibits abnormal prion protein formation in prion-infected cells. Journal of Virology. 2017;**91**(6):e01862-e01816. DOI: 10.1128/JVI.01862-16

[27] Herrera AS, Solis Arias PE. Einstein cosmological constant, the cell, and the intrinsic property of melanin to split and re-form the water molecule. MOJ Cell Science Report. 2014;**1**(2):46-51. DOI: 10.15406/mojcsr.2014.01.00011

[28] West JB. Three classical papers in respiratory physiology by Christian Bohr (1855-1911) whose work is frequently cited but seldom read. American Journal of Physiology. Lung Cellular and Molecular Physiology. 2019;**316**(4): L585-L588. DOI: 10.1152/ajplung.00527.2018 Epub 2019 Jan 17

[29] Korf J. Quantum and multidimensional explanations in a neurobiological context of mind. Neuroscientist. Aug 2015;**21**(4):345-355. DOI: 10.1177/1073858414542250. Epub 2014 Jul 21. PMID: 25049264

[30] Simeonov PL, Matsuno K, Root-Bernstein RS. Progress in biophysics and molecular biology. Can biology create a profoundly new mathematics and computation? Special theme issue on integral biomathics. Editorial. Progress in Biophysics and Molecular Biology. Sep 2013;**113**(1):1-4. DOI: 10.1016/j.pbiomolbio.2013.03.001. Epub 2013 Apr 1. PMID: 23558022

[31] Shojaie F. Quantum mechanics and the mechanism of sexual reproduction. International Journal of Advanced Biological and Biomedical Research. 2013;**1**(issue 10):1157e1164

[32] Lee H, Cheng Y-C, Fleming GR. Coherence dynamics in photosynthesis: Protein protection of exitonic coherence. Science. 2007;**316**:1462e1465

[33] Dorfman KE, Voronine DV, Mukamel S, Scully MO. Photosynthetic reaction center as a quantum heat engine. PNAS. 2013;**110**(8):2746-2751

[34] Ben-Naim A. The Protein Folding Problem and its Solutions. Singapore: World Scientific Publishing; 2013

[35] Levinthal C. Are there pathways of protein folding. Journal de chimie physique. 1968;**65**:44

[36] Berezovsky IN, Trifonov EN. Loop fold structure of proteins: Resolution of Levinthal's paradox. Journal of Biomolecular Structure & Dynamics. 2002;**20**(1):5e6

[37] Bryngelson JD, Wolynes PG. Spin glasses and the statistical mechanics of protein folding. PNAS. 1987;**84**: 7524e7528

[38] Dawkins R. The Blind Watchmaker. New York: Penguin; 1986

[39] Melkikh AV. Nonlinearity of quantum mechanics and the solution of the problem of wave function collapse. Communications in Theoretical Physics. 2015b;**64**(issue 1):47e53

[40] Melkikh AV, Seleznev VD. Mechanisms and models of the active transport of ions and the transformation of energy in intracellular compartments. Progress in Biophysics and Molecular Biology. 2012;**109**(Issues 1e2):33e57

[41] Allemann RK, Scrutton NS, editors. Quantum Tunnelling in Enzyme-catalysed Reactions RSC Biomolecular Sciences. Royal Society of Chemistry (Great Britain). 2009

[42] Lloyd S. A quantum of natural selection. Nature. 2009;**5**:164e166

[43] Baez JC. Is life improbable? Foundations of Physics. 1989;**19**(1):91e95

[44] Freeland SJ, Hurst LD. The genetic code is one in a million. Journal of Molecular Evolution. 1998;**47**:238e248

[45] Schoels GD. Quantum coherent electronic energy transfer: Did nature think of it first? Journal of Physical Chemistry Letters. 2010;**1**(1):2e8

[46] Derek A et al. Quantum Aspects of Life. World Scientific Publishing Co. Pte. Ltd.; 2008. pp. 223-231. ISBN #9781848162556

[47] Trevors JT, Masson L. Quantum microbiology. Current Issues in Molecular Biology. 2012;**13**:43e50

[48] Asano M, Basieva I, Khrennikov A, Ohya M, Tanaka Y, Yamato I. A model of epigenetic evolution based on theory of open quantum systems. Systems and Synthetic Biology. 2013b;**7**:161e173

[49] Tarlacı S, Pregnolato M. Quantum neurophysics: From non-living matter to quantum neurobiology and psychopathology. International Journal of Psychophysiology. May 2016;**103**:161-173. DOI: 10.1016/j.ijpsycho.2015.02.016. Epub 2015 Feb 7. PMID: 25668717

[50] Sharp D. John Scott Haldane, respiration and public health. The Lancet. 2008;**372**:795-796

Chapter 6

Reproductive Strategies of the Female Gametophyte

Xiaoyan Liu and Ryushiro D. Kasahara

Abstract

Reproduction refers to the biological process of producing offspring. Sexual reproduction in angiosperms is a complex and precise process of regulation, which requires the mutual recognition of male and female gametes. The pollen tube, as a medium for transmission of male gametes, is attracted by chemoattractant derived from synergid cells in a target ovule. We first identified that MYB98 plays an important role in pollen tube guidance by regulating the downstream LURE peptides. Moreover, we discovered that if the first pollen tube penetrates the ovule but fertilization fails, the second pollen tube is attracted by another synergid cell to increase the chances of fertilization as a "wise" back-up system (Fertilization Recovery System). Similar feedback mechanisms also occur to seed development after the pollen tube guidance. We further identified a phenomenon, the pollen tube-dependent ovule enlargement morphology (POEM), that the pollen tube contents (PTCs) enlarge ovules and initiate seed coat formation without fertilization. Furthermore, we identified the POEM in rice by knocked-out GCS1 genes in rice genome, which led to fertilization failure and to produce enlarged sugar grain. In this chapter, we discuss from the reproductive strategies of the plants to the agricultural application based on our previous discoveries.

Keywords: pollen tube guidance, MYB98, LUREs, fertilization recovery system, POEM, sugar rice

1. Introduction

Seed plants, also known as higher vascular plants, include gymnosperms and angiosperms, which have evolved a unique life cycle: flowering, pollination, fertilization, and seed formation. Double fertilization is a flowering plant mechanism whereby two immotile sperm cells fertilize two different female gametes. In order to achieve the fusion of sexual gametes (fertilization), the mature pollen (male gametophyte) is dispersed from the anther to the stigmas that undergo the long journey to the ovule (inside the ovule is the female gametophyte), which requires the mutual recognition of male and female gametes (**Figure 1**). In this chapter, we will discuss from the pollen tube guidance, fertilization, the induction of seed development, and to the agricultural application based on our previous discoveries.

IntechOpen

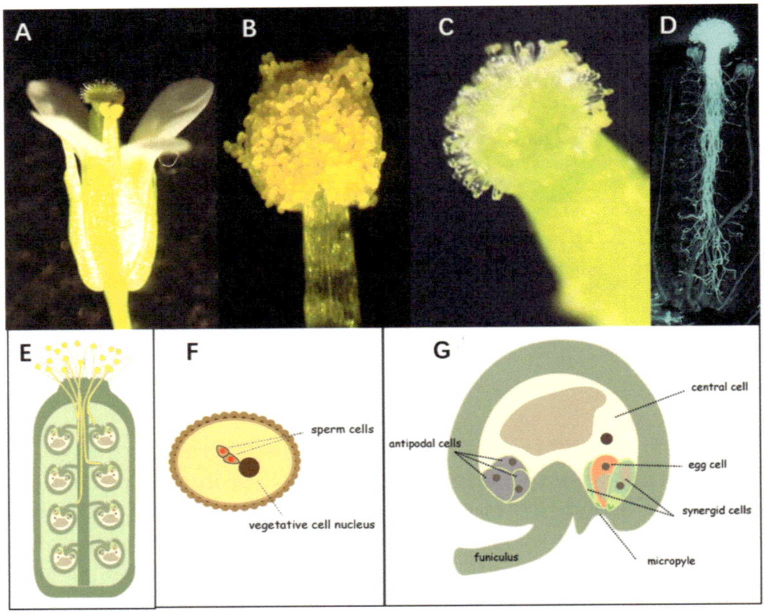

Figure 1.
The Arabidopsis male and female gametophyte. A. The flower of Arabidopsis. B. Anther of Arabidopsis thaliana, Anthers dehiscent, exposing pollen grains. C. Stigmas bearing pollen grains. D. Pollen tube stained by aniline blue, WT self-cross 8HAP. E. Pollens fall onto the stigma to grow pollen tubes. F. Male gametophyte (pollen grain), with one vegetative cell and two sperm cells inside. G. Female gametophyte, consists of seven cells and four different cell types: three antipodal cells, two synergid cells, one egg cell, and one central cell.

2. Development of gametophytes

During the angiosperm life cycle, the angiosperm gametophytes develop within sporophytic tissues that constitute the sexual organs of the flower. The sporophyte produces two types of spores, microspores and megaspores, that give rise to male gametophytes and female gametophytes, respectively [1]. Female gametophyte development goes through two processes referred to as megasporogenesis and megagametogenesis. The diploid megaspore mother cell undergoes meiosis and gives rise to four haploid megaspores. After that, three megaspores go through cell death, and the only megaspore left goes through three rounds of mitosis without cytokinesis, resulting in a multinucleate coenocyte. Subsequently, cell walls form around these nuclei, forming the cellular female gametophyte (**Figure 1G**), which consists of seven cells and four different cell types: three antipodal cells, two synergid cells, one egg cell, and one central cell [2]. The male gametophyte (**Figure 1F**), also referred to as the pollen grain or microgametophyte, develops within the anther (**Figure 1B**) and is composed of two sperm cells encased within a vegetative cell.

3. Pollen adhesion

Once a pollen grain adheres to the stigma, compatible pollen must be distinguished from potential pathogens (fungal spores or bacteria). In response to compatible pollen, stigmas initiate a basal response pathway that transfers water to the desiccated pollen grain for pollen hydration and germination [3].

In self-incompatible plants of the genus Brassica, self-related pollen grains are recognized and prevented to germinate by interaction with the epidermal cells of the stigma (**Figure 2B**). The self-incompatibility (SI) phenomenon is genetically controlled by a complex and polymorphic locus [4]. Among the genes, S locus is a pair of sequence-related genes, the cell wall localized S-locus glycoprotein (*SLG*) gene and the plasma membrane spanning receptor protein kinase (*SRK*) gene, both of which are expressed specifically in the stigma epidermal cells. The pollen coat localized S-locus cysteine-rich/S-locus protein 11 (SCR/SP11) ligands interacts with S-receptor kinase (SRK), thereby inducing its self-activation during the self-pollen recognition process [5–7]. However, for compatible pollen, due to the absence of pollen coat protein, self-activation of SRK is inhibited by thioredoxin H-like1 (THL1), resulting pollen hydration and germination [8, 9]. Upon phosphorylation and activation, SRK forms a complex with M-locus protein kinase (MLPK, a plasma-membrane-localized receptor-like cytoplasmic kinase protein), which in turn interacts with arm-repeat-containing protein 1 (ARC1, a pistil specific E3 ubiquitin ligase protein) [10, 11]. The activated ARC1 is directly involved in the ubiquitination of Exo70A1, a plasma membrane-localized peptide, which is involved in the secretory vesicle delivery to the papillae surface [12–14]. The study shows that in the absence of functional Exo70A1, the exocyst complex-derived stigmatic secretion failed [15]. In addition, studies indicate that pollen-specific plasma-membrane-localized aquaporins (NIP4;1,

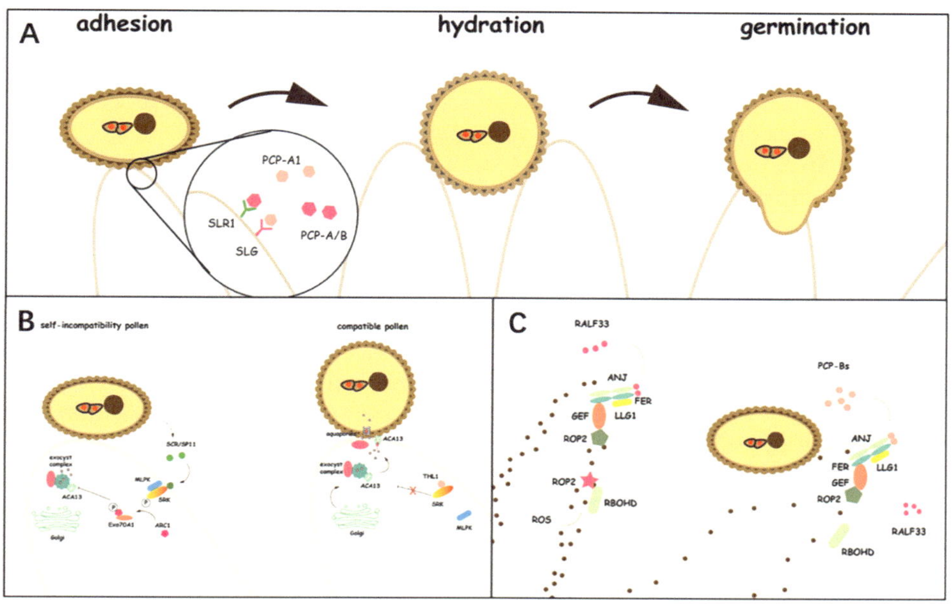

Figure 2.
Pollen landing on stigma to germinate pollen tube. A. Pollen adhesion, hydration, and germination. B. SRK recognizes pollen of the same plant by interacting with SCR/SP11 thereby phosphorylating ARC1 which, in turn, interacts with Exo70A1 and evokes self-incompatibility by hindering stigmatic secretion. In the absence of SRK interaction with pollen-specific SCR/SP11, it interacts with stigma-specific THL1 and ARC1 phosphorylation is hindered. Hence, normal stigmatic secretion proceeds and pollen gets hydrated. C. Model of compatible pollen recognition. Before pollination, RALF23/33 induces ROS production in the stigmatic papilla cells through an ANJ-FER-ROP2-RBOHD pathway. After pollination with compatible pollen, PCP-Bs from the pollen coat compete with RALF23/33 for interaction with the ANJ-FER complex, repressing ROS production and initiating stigmatic responses.

NIP4;2, TIP1;3, TIP5;1) are potential macromolecules involved in pollen hydration by rendering pollen PM permeable to water and other solutes [16, 17].

Although the main mechanisms of action of different types of self-incompatibility systems have been clarified, in contrast, relatively little attention has been paid to compatible pollen recognition and the corresponding cellular responses in the stigmatic papillae. In *Arabidopsis thaliana*, the pollen coat PCP-Bs are excellent candidates for the pollen ligand to initiate the basal compatible pollen acceptance pathway in the papilla. The loss of PCP-Bs notably slows pollen hydration and germination [18]. Recently, a stigmatic gatekeeper, the ANJEA–FERONIA (ANJ–FER) receptor kinase complex, perceives the RAPID ALKALINIZATION FACTOR peptides RALF23 and RALF33 to induce reactive oxygen species (ROS) production in the stigma papillae, whereas pollination reduces stigmatic ROS, allowing pollen hydration. Upon pollination, PCP-Bs compete with RALF23/33 for binding to the ANJ-FER complex, leading to a decline of stigmatic ROS that facilitates pollen hydration (**Figure 1C**) [19].

4. Pollen tube guidance at the micropylar region

Several studies have reported that the pollen tubes fail to grow onto ovules containing abnormal female gametophytes, suggesting that the embryo sac provides a guiding cue for the pollen tubes [20–22]. The study on Oenothera showed that PT directly enters synergid cell, which is where it bursts and releases their contents along with two sperm cells. Later studies further demonstrated the synergid cell as being the only source of short-distance pollen tube attractants at the micropylar region [23, 24]. The synergid cells have structural specializations that facilitate the fertilization process. Each synergid cell consists of large vacuole at its chalazal end, and a nucleus and stack of ER, along with Golgi complexes, at its micropylar end [25]. The micropylar tip of each synergid cell wall is extensively invaginated, forming a structure referred to as the filiform apparatus [26]. MYB98, a synergid cell-specific R2R3 transcription factor, affects the development of filiform apparatus and is crucially important for micropylar PT guidance [27].

A later study showed that MYB98 affects the expression of arrays of SC-specific genes encoding defensin-like cysteine-rich proteins (CRPs), which are secreted into the filiform apparatus and are involved in PT guidance [28]. LURE1 and LURE2, two of the synergid cell produced CRPs, are directly involved in micropylar pollen tube guidance [29, 30]. Recent studies have shown that these proteins interact with receptors such as PRK6 [31] and MIDIS1-MIK [32] that are produced by the PTs and guide it toward the micropylar. The pollen tube tip-specific LIP1 and LIP2, members of RLKs, also regulate micropylar PT guidance and affect PT attraction toward LURE1 (**Figure 3**A). The study in *Torenia fournieri* showed that AMOR, an arabinogalactan polysaccharide secreted by mature ovules, also plays a positive role in rendering the growing PT competent to interact with LURE attractants [33]. The AtLURE1/PRK6-mediated signaling pathway thus guarantees a strong and biased precedence for own pollen tubes and contributes to prezygotic reproductive isolation in the genus Arabidopsis. Except LUREs, another group of CRPs, XIUQIU1–4 (**Figure 3A**), attracts PTs without any species bias [34, 35]. However, the interaction of XIUQIU to as yet unknown PT-receptor/s is likely conserved across Brassicaceae members as it can effectively attract PTs of *Arabidopsis lyrata* and Ca*ps*ella *rubella*.

In addition to the important role of synergid cell in pollen tube guidance, central cells also seem to influence pollen tube guidance. A central cell-specific Central Cell Guidance

DOI: http://dx.doi.org/10.5772/intechopen.109805

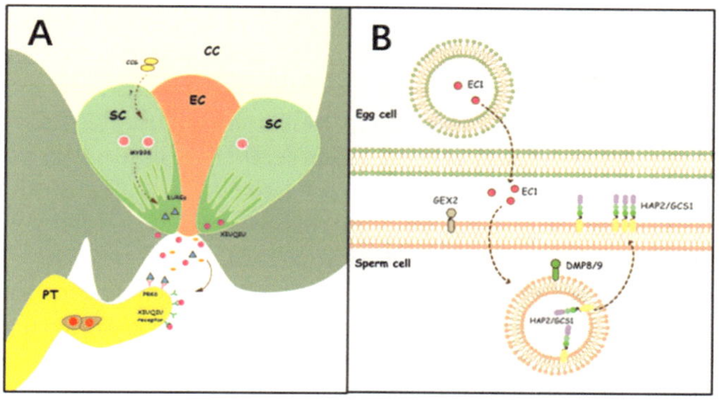

Figure 3.
Pollen tube guidance and double fertilization. A. Micropylar guidance is regulated by LURE-PRK6/MDIS1-MIK interaction, TIC2, XIUQIU, AMOR. Central cell localized CCG and synergid cell localized MYB98 affect CRP biosynthesis. B. HAP2/GCS1 interacts with DMP8 and DMP9, which are required for the EC1-induced translocation of HAP2/GCS1 from internal storage vesicle to the sperm plasma membrane to ensure successful fertilization.

(CCG), encoding a nuclear protein with an N-terminal conserved zinc b-ribbon domain, has also been reported to affect micropylar pollen tube guidance [36, 37]. Interestingly, in *ccg* mutant, MYB98 and other CRP genes, including LURE1, are downregulated [37]. This is most likely due to the changes in MYB98 expression, or MYB98 along with CCG and CBP1, co-regulate the expression of those CRPs. However, if they were to co-regulate along with MYB98, additional evidence and explanations are required [38].

5. Discharge of sperm cells from the pollen tube tip to fertilization

Pollen tube growth arrests within a synergid and the pollen tube tip bursts, releasing the PT contents, including the two sperm cells [39]. This process is accompanied by the degeneration of the receptive synergid and is rapidly followed by the fusion of one sperm with the egg and one sperm with the central cell to give rise to the zygote and endosperm, respectively [40, 41].

Double fertilization proceeds through the recognition, attachment, and membrane fusion of male and female gametes, and these processes are directly regulated by proteins on the gamete surface (**Figure 3B**) [42]. Male gametic membrane proteins HAP2/GCS1, GEX2, DMP8, and DMP9 have been identified as fertilization regulators affecting male-female gamete fusion [43–46]. The first gamete fusogen identified was HAP2/GCS1, which share a similar structure with two distinct families of exoplasmic fusogens: the somatic Fusion Family (FF) and class II viral glycoproteins [47]. The egg cell-specific EC1, which was reported to be crucial for sperm cell activation [48], also plays an important role in sperm cell attachment to the plasma membrane of both the egg cell and central cell [46]. EC1 proteins accumulate in storage vesicles of the egg cell. Upon sperm arrival, EC1-containing vesicles are exocytosed. The sperm endomembrane system responds to exogenously applied EC1 peptides by redistributing the potential gamete fusogen HAP2/GCS1 to the cell surface. These findings provide evidence that mutual gamete activation, regulated exocytosis, and sperm plasma membrane modifications govern flowering plant gamete interactions. However, it is yet unclear whether the EC1 acts solely or in coordination with sperm cell-specific signal during the process. Recent study reported that two sperm DUF679 membrane proteins DMP8 and DMP9 interact with HAP2/GCS1, which are required for the EC1-induced translocation of HAP2/GCS1 from internal storage vesicle to the sperm plasma membrane to ensure successful fertilization [49].

Once these two cells fuse at fertilization, their nuclei must then navigate toward each other and fuse. When an animal egg cell is fertilized, cable-like protein filaments called microtubules guide the two nuclei into contact. These microtubules are organized by a cellular structure called a centrosome [50, 51]. However, flowering plants do not have centrosomes [52]. Kawashima et al. [53] found that the fertilization requires an intact F-actin network. The sperm nucleus becomes surrounded by a star-shaped structure of F-actin cables, and that this F-actin structure migrates together with the sperm nucleus. The F-actin network constantly moves inward, from the edges of the cell toward the nucleus, prior to fertilization. ROP8 is a female gamete-specific Rho-GTPase that regulates F-actin dynamics. Previous research has shown that the Wiskott-Aldrich syndrome protein family verprolin-homologous and suppressor of the cAMP receptor (WAVE/SCAR) family are effector proteins that directly interact with ROPs and promote actin nucleation [54, 55]. The WAVE/SCAR complex is the main activator of the F-actin regulatory ACTIN RELATED PROTEIN 2/3 (ARP2/3) protein complex [56, 57]. However, further study demonstrated that the F-actin regulator, SCAR2, but not the ARP2/3 protein complex, regulates F-actin dynamics in female gametophytic cells for fertilization. In addition, the class XI myosin XI-G controls active F-actin movement in the Arabidopsis central cell [58].

6. Fertilization recovery system and polytubey blocking

In normal conditions, although hundreds of pollens may land on the stigma and growing pollen tubes into the transmitting tract of a pistil, usually only a single tube, in response to attractants, emerges from the septum in the vicinity of each ovule to target the ovule [59, 60]. Even the attraction of more than one pollen tube has been observed in different gametophytic mutants [61–65]. The mechanism regulating pollen tube number remains unclear. As we discussed earlier, the synergid plays a role in attracting the pollen tube. In Arabidopsis, pollen tube growth arrests within a synergid cell and accompanied by the degeneration of the receptive synergid. Thus, the second pollen tube is attracted by another synergid. Kasahara et al. [66, 67] investigated the mechanisms underlying this phenomenon in Arabidopsis upon frequently observing ovules that accepted two pollen tubes in the fertilization defective *hap2–1* (allelic to *gcs1*) mutant. They observed that the fertility rate of the ovules pollinated with pollen from male gametophytic mutants was not around 50%, as expected, it is 60–70%. This result was revealed once fertilization failed with the first pollen tube, attracted a second one through a second synergid cell, so increasing the fertility rate. This phenomenon is called fertilization recovery system (**Figure 4A** and **B**).

In addition to the fertilization recovery system, plants, like animals, also have an important task of avoiding polyspermy. In Arabidopsis, the block to polyspermy is facilitated by a mechanism that prevents polytubey. How exactly do plants control the number of pollen tubes they attract? Three block mechanisms have been discovered recently, occurring in the septum, the micropyle, and after fertilization. The first polytubey block is located at the septum, FERONIA, ANJEA, and HERCULES RECEPTOR KINASE 1 receptor-like kinases located at the septum interact with pollen tube-specific RALF6, 7, 16, 36, and 37 peptide ligands to establish polytubey block (**Figure 4D**) [68]. In addition, Duan et al. [69] demonstrated that pollen tube arrival at

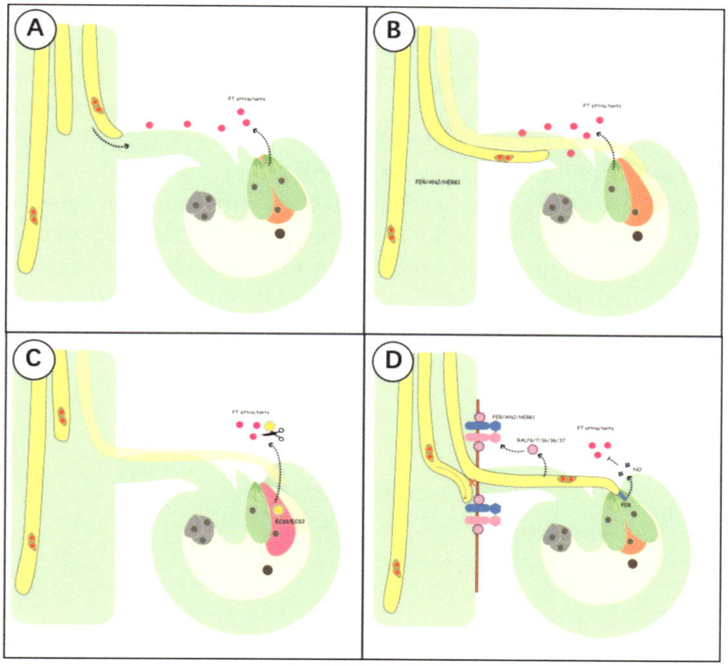

Figure 4.
Fertilization recovery system and polytubey. A. One pollen is being directed toward the ovule. B. The first pollen tube penetrates the ovule but fertilization fails, the second pollen tube is attracted by another synergid cell to increase the chances of fertilization as a "wise" back-up system (Fertilization Recovery System). C. Fertilization signals (ECS1 and ECS2) block polytubey. ECS1 and ESC2 specifically cleave the pollen tube attractor LURE1. D. FERONIA, ANJEA, and HERCULES RECEPTOR KINASE 1 receptor-like kinases located at the septum interact with pollen tube-specific RALF6, 7, 16, 36, and 37 peptide ligands to establish polytubey block. In addition, NO affects LURE1 and suppresses pollen tube attraction at micropylar.

the ovule triggers the accumulation of nitric oxide at the filiform apparatus in a process that is dependent on FERONIA and mediated by de-esterified pectin. Nitric oxide nitrosates both precursor and mature forms of the chemoattractant LURE11, respectively, blocking its secretion and interaction with its receptor, to suppress pollen tube attraction. Fertilization signals can also block polytubey. After successful fertilization, the aspartic endopeptidases ECS1 and ECS2 are secreted to the extracellular space from a cortical network located at the apical domain of the Arabidopsis egg cell. ECS1 and ESC2 specifically cleave the pollen tube attractor LURE1 (**Figure 4C**). In consequence, polytubey is frequent in *ecs1 ecs2* double mutants. These findings demonstrate that plant egg cells sense successful fertilization and elucidate a mechanism as to how a relatively fast post-fertilization block to polytubey is established by fertilization-induced degradation of attraction factors [70].

7. Pollen tube-dependent ovule enlargement morphology (POEM)

As discussed earlier, once two sperm cells fused to egg cell and central cell, to give rise to embryo and endosperm, which is the start of seed development. However, nobody knows the mechanism of initiation of seed development, except double fertilization. The discovery of pollen tube-dependent ovule enlargement morphology (POEM) brings a turning point, interesting in that the PTC itself is enough to initiate the ovule enlargement (**Figure 5**) even though it may not grow to the regular seed size [71]. As a preparation for fertilization, PTC initiated the developmental activity of various cells in the ovule. While studies have suggested that fertilization is a prerequisite for seed coat development [72, 73]. The studies demonstrated that just PTC release into the ovule is enough to initiate seed coat and endosperm development [74, 75]. In addition, the expression of numerous genes is responsible for cell expansion, cell division, and seed coat development in *gcs1/gcs1* pollinated ovules. As a preparation for fertilization, PTC initiates the developmental activity of various cells in the ovule. After mobilization, the ovules are waiting for the signal from double fertilization to continue the seed development, in addition to the phenotype of ovule size and seed coat formation.

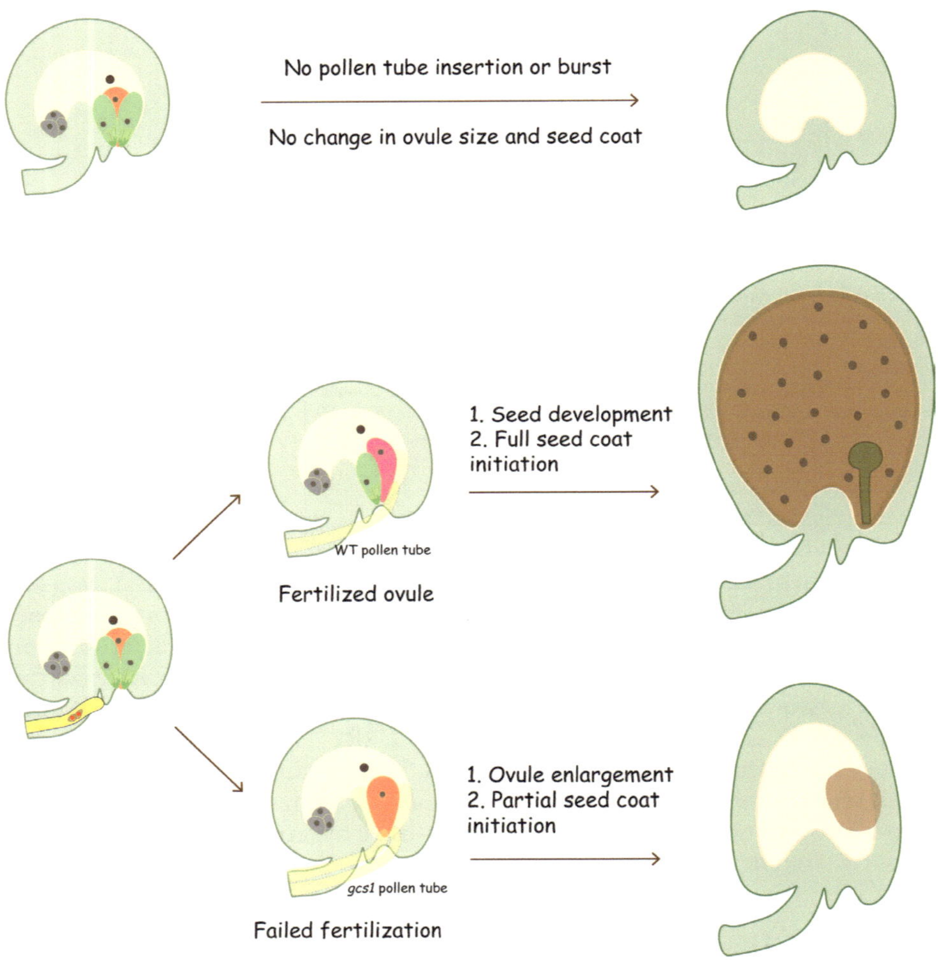

Figure 5.
Pollen tube-dependent ovule enlarged morphology (POEM). After the pollen tube arrests within a synergid, the pollen tube bursts and releases its contents with two sperm cells. Double fertilization is accomplished by these sperm cells fertilizing egg cell and central cell to give rise to the zygote and endosperm, respectively. However, if the ovule gets gcs1 mutant pollen tube, gcs1 sperm cells fail to fertilize. The ovule will be enlarged and initiate seed coat formation without fertilization.

Discovery of the POEM phenomenon showed its potential applications in crop breeding for seed size increment and apomixis induction. However, POEM has only been reported in Arabidopsis. There are great differences between mono and dicotyledon plants. To investigate if this phenomenon is conserved in monocot as well, Honma et al. [76] developed genome-edited rice plants by knocking-out homologs of rice *GCS1* using the CRISPR/Cas9 technology, which led to fertilization failure and pollen tube-dependent ovule enlargement morphology (POEM) phenomenon. Apparently, the POEMed-like rice ovule can grow near-normal seed size unlike earlier observations in Arabidopsis in which *gcs1* ovules were aborted quite early. The POEMed-like rice ovules contained 10–20% sugar, with extremely high sucrose content (98%). Transcriptomic analysis revealed that the osgcs1 ovules had downregulation of starch biosynthetic genes, which would otherwise have converted sucrose to starch. Overall, this study shows that pollen tube content release is sufficient to trigger sucrose unloading at rice ovules. However, successful fertilization is indispensable to trigger sucrose-starch conversion. These findings are expected to pave the way for developing novel sugar-producing crops suited for diverse climatic regions.

8. Summary

This chapter discusses the journey of the pollen from the stigma to fertilization as well as the POEM phenomenon. These processes are achieved through the involvement of various male-female interactions. The molecular mechanisms underlying pollen tube guidance from the funiculus to the female gametophyte are well known in Arabidopsis, because the pollen tube attractants AtLURE1 peptides had previously been identified downstream of the master synergid cell regulator MYB98. Recently, more and more factors related to fertilization journey have been elucidated. During the final step after pollen tube bursting, HAP2/ GCS1, GEX2, DMP8, and DMP9 have been identified as direct male-related key fertilization factors. The F-actin network is also required for sperm nucleus migration. The fertilization recovery system is to increase the chances of fertilization as a "wise" back-up system. Combining fertilization recovery system with recent research on polytubey blocking, the process of how ovule attracts pollen tube is becoming clear. Finally, very few factors related to new plant phenomena and POEM have been identified. New insights into the underlying molecular mechanisms are anticipated.

Acknowledgements

These works were supported by the Precursory Research for Embryonic Science and Technology (13416724, Kasahara Sakigake Project), Japan Science and Technology Agency. These works were also supported by a grant-in-aid (25840106) from the Japanese Society for the promotion of Science (JSPS). These works were also supported by FAFU-UCR Joint Center and Fujian Provincial Key Laboratory of Haixia Applied Plant Systems Biology, Haixia Institute of Science and Technology, Fujian Agriculture and Forestry University.

Conflict of interest

The authors declare no conflict of interest.

Author details

Xiaoyan Liu[1,2]* and Ryushiro D. Kasahara[1,2,3]

1 School of Life Sciences, Fujian Agriculture and Forestry University, Fuzhou, China

2 FAFU-UCR Joint Center and Fujian Provincial Key Laboratory of Haixia Applied Plant Systems Biology, Haixia Institute of Science and Technology, Fujian Agriculture and Forestry University, Fuzhou, China

3 Graduate School of Science and Technology, Kumamoto University, Kumamoto, Japan

*Address all correspondence to: liuxy@fafu.edu.cn

IntechOpen

© 2023 The Author(s). Licensee IntechOpen. This chapter is distributed under the terms of the Creative Commons Attribution License (http://creativecommons.org/licenses/by/3.0), which permits unrestricted use, distribution, and reproduction in any medium, provided the original work is properly cited.

References

[1] Twell D. Male gametogenesis and germline specification in flowering plants. Sexual Plant Reproduction. 2011;**24**:149-160. DOI: 10.1007/s00497-010-0157-5

[2] Yadegari R. Female gametophyte development. The Plant Cell. 2004;**16**:S133-S141. DOI: 10.1105/tpc.018192

[3] Jennifer, Doucet, Hyun, et al. Pollen acceptance or rejection: A tale of two pathways. Trends in Plant Science. 2016;**21**:1058-1067. DOI: https://doi.org/10.1016/j.tplants

[4] Boyes DC, Nasrallah ME, Vrebalov J, et al. The self-incompatibility (S) haplotypes of brassica contain highly divergent and rearranged sequences of ancient origin. The Plant Cell. 1997;**9**:237-247. DOI: 10.1105/tpc.9.2.237

[5] Schopfer CR, Nasrallah ME, Nasrallah JB, Schopfer CR, Nasrallah ME, Nasrallah JB. The male determinant of self-incompatibility in brassica. Science. 1999;**286**:1697-1700. DOI: 10.1126/science.286.5445.1697

[6] Kemp BP, Doughty J. S cysteine-rich (SCR) binding domain analysis of the brassica self-incompatibility S-locus receptor kinase. The New Phytologist. 2007;**175**:619-629. DOI: 10.111 1/j.1469-8137.2007.02126.x

[7] Shimosato H, Yokota N, Shiba H, Iwano M, Entani T, Che F-S, et al. Characterization of the SP11/SCR high-affinity binding site involved in self/nonself recognition in Brassica self-incompatibility. The Plant Cell. 2007;**19**:107-117. DOI: 10.1105/tpc.105.038869

[8] Bower MS, Matias DD, Fernandes-Carvalho E, Mazzurco M, Gu T, Rothstein SJ, et al. Two members of the thioredoxin-h family interact with the kinase domain of a Brassica S locus receptor kinase. The Plant Cell. 1996;**8**:1641-1650. DOI: 10.1105/tpc.8.9.1641

[9] Cabrillac D, Cock JM, Dumas C, Gaude T. The S-locus receptor kinase is inhibited by thioredoxins and activated by pollen coat proteins. Nature. 2001;**410**:220-223. DOI: 10.1038/35065626

[10] Gu T, Mazzurco M, Sulaman W, Matias DD, Goring DR. Binding of an arm repeat protein to the kinase domain of the S-locus receptor kinase. Proceedings of the National Academy of Sciences. 1998;**95**:382-387. DOI: 10.1073/pnas.95.1.382

[11] Stone SL, Anderson EM, Mullen RT, Goring DR. ARC1 is an E3 ubiquitin ligase and promotes the ubiquitination of proteins during the rejection of self-incompatible Brassica pollen. The Plant Cell. 2003;**15**:885-898. DOI: 10.1105/tpc.009845

[12] Samuel MA, Chong YT, Haasen KE, Aldea-Brydges MG, Stone SL, Goring DR. Cellular pathways regulating responses to compatible and self-incompatible pollen in Brassica and Arabidopsis stigmas intersect at Exo70A1, a putative component of the exocyst complex. The Plant Cell. 2009;**21**:2655-2671. DOI: 10.1105/tpc.109.069740

[13] Yang K, Zang HC, Converse R, Zhu LQ, Yang YJ, Xue LY, et al. Interaction between two self-incompatible signal elements, EXO70A1 and ARC1. Acta Agronomica Sinica. 2012;**37**:2136-2144. DOI: 10.1016/S1875 -2780(11)60054-0

[14] Liu J, Zhang H, Lian X, Converse R, Zhu L. Identification of interacting motifs between armadillo repeat containing 1 (ARC1) and exocyst 70 A1 (Exo70A1) proteins in *Brassica oleracea*. The Protein Journal. 2016;**35**:34-43. DOI: 10.1007/s1093 0-015-9644-8

[15] Safavian D, Goring DR. Secretory activity is rapidly induced in stigmatic papillae by compatible pollen, but inhibited for self-incompatible pollen in the Brassicaceae. PLoS One. 2013;**8**:e84286. DOI: 10.1371/journal.pone.0084286

[16] Wudick MM, Luu DT, Tournaire Roux C, Sakamoto W, Maurel C. Vegetative and sperm cell-specific aquaporins of Arabidopsis highlight the vacuolar equipment of pollen and contribute to plant reproduction. Plant Physiology. 2014;**164**:1697-1706. DOI: 10.1104/pp.113.228700

[17] Pérez DGJA, Barberini ML, Amodeo G, Muschietti JP. Pollen aquaporins: What are they there for? Plant Signaling & Behavior. 2016;**11**:e1217375-e1217375. DOI: 10.1080/15592 324.2016.1217375

[18] Wang L, Clarke LA, Eason RJ, Parker CC, Qi B, Scott RJ, et al. PCP-B class pollen coat proteins are key regulators of the hydration checkpoint in *Arabidopsis thaliana* pollen-stigma interactions. The New Phytologist. 2017;**213**:764-777

[19] Liu C, Shen L, Xiao Y, et al. Pollen PCP-B peptides unlock a stigma peptide-receptor kinase gating mechanism for pollination. Science. 2021;**372**:171-175. DOI: 10.1126/science.abc6107

[20] Hulskamp M, Schneiz K, Pruit RE. Genetic evidence for a long-range activity that directs pollen tube guidance in Arabidopsis. The Plant Cell. 1995;7:57-64

[21] Ray A. Three's company: Regulatory cross-talk during seed development. The Plant Cell. 1997;**9**:665-667

[22] Shimizu KK, Attractive OK. Repulsive interactions between female and male gametophytes in Arabidopsis pollen tube guidance. Development. 2000;**127**:4511-4518

[23] Higashiyama T, Kuroiwa H, Kawano S, Kuroiwa T. Guidance in vitro of the pollen tube to the naked embryo sac of *Torenia fournieri*. The Plant Cell. 1998;**10**:2019-2031. DOI: 10.1105/tpc.10.12.2019

[24] Higashiyama T, Yabe S, Sasaki N, Nishimura Y, Miyagishima SY, Kuroiwa H, et al. Pollen tube attraction by the synergid cell. Science. 2001;**293**:1480-1483. DOI: 10.1126/science.1062429

[25] Jensen WA. The ultrastructure and histochemistry of the synergids of cotton. American Journal of Botany. 1965;**52**:238-256. DOI: 10.1002/j.1537-2197.1965.tb06781.x

[26] Gunning BES, Pate JS. "Transfer cells" plant cells with wall ingrowths, specialized in relation to short distance transport of solutes—Their occurrence, structure, and development. Protoplasma. 1969;**68**:107-133. DOI: 10.1007/bf01247900

[27] Kasahara RD, Portereiko MF, Sandaklie Nikolova L, Rabiger DS, Drews GN. MYB98 is required for pollen tube guidance and synergid cell differentiation in Arabidopsis. The Plant Cell. 2005;**17**(11):2981-2992. DOI: 10.1105/tpc.105.034603

[28] Punwani JA, Rabiger DS, Drews GN. MYB98 positively regulates a battery of synergid-expressed genes encoding filiform apparatus–localized proteins.

The Plant Cell. 2007;**19**:2557-2568. DOI: 10.1105/tpc.107.052076

[29] Okuda S, Tsutsui H, Shiina K, Sprunck S, Takeuchi H, Yui R, et al. Defensin-like polypeptide LUREs are pollen tube attractants secreted from synergid cells. Nature. 2009;**458**:357. DOI: 10.1038/nature07882

[30] Takeuchi H, Higashiyama T. A species-specific cluster of defensin-like genes encodes diffusible pollen tube attractants in Arabidopsis. PLoS Biology. 2012;**10**:e1001449. DOI: 10.1371/ journal.pbio.1001449

[31] Takeuchi H, Higashiyama T. Tip-localized receptors control pollen tube growth and LURE sensing in Arabidopsis. Nature. 2016;**531**:245. DOI: 10.1038/nature17413

[32] Wang T, Liang L, Xue Y, Jia PF, Chen W, Zhang MX, et al. A receptor heteromer mediates the male perception of female attractants in plants. Nature. 2016;**531**:241. DOI: 10.1038/nature16975

[33] Mizukami Akane G, Inatsugi R, Jiao J, Kotake T, Kuwata K, Ootani K, et al. The AMOR arabinogalactan sugar chain induces pollen-tube competency to respond to ovular guidance. Current Biology. 2016;**26**:1091-1097. DOI: 10.1016/j.cub.2016.02.040

[34] Meng JG, Zhang MX, Yang WC, Li HJ. TICKET attracts pollen tubes and mediates reproductive isolation between relative species in Brassicaceae. Science China. Life Sciences. 2019;**62**:1413-1419. DOI: 10.1007/s11427-019-9833-3

[35] Zhong S, Liu M, Wang Z, Huang Q, Hou S, Xu YC, et al. Cysteine-rich peptides promote interspecific genetic isolation in Arabidopsis. Science. 2019;**364**:9564. DOI: 10.1126/science.aau9564

[36] Chen YH, Li HJ, Shi DQ, Yuan L, Liu J, Sreenivasan R, et al. The central cell plays a critical role in pollen tube guidance in Arabidopsis. The Plant Cell. 2007;**19**:3563-3577. DOI: 10.1105/tpc.107.053967

[37] Li HJ, Zhu SS, Zhang MX, Wang T, Liang L, Xue Y, et al. Arabidopsis CBP1 is a novel regulator of transcription initiation in central cell-mediated pollen tube guidance. The Plant Cell. 2015;**27**:2880-2893. DOI: 10.1105/tpc.15.00370

[38] Erdmann RM, Hofmann A, Walter H-K, Wagenknecht H-A, GroßHardt R, Gehring M. Molecular movement in the *Arabidopsis thaliana* female gametophyte. Plant Reproduction. 2017;**30**:141-146. DOI: 10.1007/s00497-017-0304-3

[39] Russell S. Double fertilization. International Review of Cytology. 1992;**140**:357-388

[40] Faure JE, Dumas C. Fertilization in flowering plants. New approaches for an old story. Plant Physiology. 2001;**125**:102-104

[41] Weterings K, Russell SD. Experimental analysis of the fertilization process. The Plant Cell. 2004;**16**:S107-S118

[42] Mori T, Kawai-Toyooka H, Igawa T, Igawa T, Nozaki H. Gamete dialogs in green lineages. Molecular Plant. 2015;**8**:1442-1454

[43] Mori T, Kuroiwa H, Higashiyama T, Kuroiwa T. GENERATIVE CELL SPECIFIC 1 is essential for angiosperm fertilization. Nature Cell Biology. 2006;**8**:64-71

[44] von Besser K, Frank AC, Johnson MA, Preuss D. Arabidopsis HAP2 (GCS1) is a sperm-specific gene

required for pollen tube guidance and fertilization. Development. 2006;**133**:4761-4769. DOI: 10.1242/dev.02683

[45] Mori T, Igawa T, Tamiya G, Igawa T, Nozaki H. Gamete attachment requires GEX2 for successful fertilization in Arabidopsis. Current Biology. 2014;**24**:170-175

[46] Cyprys P, Lindemeier M, Sprunck S. Gamete fusion is facilitated by two sperm cell-expressed DUF679 membrane proteins. Nature Plants. 2019;**5**:253-257. DOI: 10.1038/s41477-019-0382-3

[47] Brukman NG, Li X, Podbilewicz B. Fusexins, HAP2/GCS1 and evolution of gamete fusion. Frontiers in Cell and Development Biology. 2022;**9**:824024. DOI: 10.3389/fcell.2021.824024

[48] Sprunck S, Rademacher S, Vogler F, Gheyselinck J, Grossniklaus U, Dresselhaus T. Egg cell-secreted EC1 triggers sperm cell activation during double fertilization. Science. 2012;**338**:1093-1097. DOI: 10.1126/science.1223944

[49] Wang W, Xiong H, Zhao P, Sun M. DMP8 and 9 regulate HAP2/GCS1 trafficking for the timely acquisition of sperm fusion competence. Proceedings of the National Academy of Sciences. 2022;**45**:119

[50] Schatten G. The centrosome and its mode of inheritance: The reduction of the centrosome during gametogenesis and its restoration during fertilization. Developmental Biology. 1994;**165**:299-335. DOI: 10.1006/dbio.1994.1256

[51] Reinsch S, Gonczy P. Mechanisms of nuclear positioning. Journal of Cell Science. 1998;**111**:2283-2295

[52] Carvalho Santos Z, Azimzadeh J, Pereira Leal JB, Bettencourt DM. Evolution: Tracing the origins of centrioles, cilia, and flagella. Journal of Cell Biology. 2011;**194**:165-175. DOI: 10.1083/jcb.201011152

[53] Kawashima T, Maruyama D, Shagirov M, Li J, Hamamura Y, Yelagandula R, et al. Dynamic F-actin movement is essential for fertilization in *Arabidopsis thaliana*. Elife. 2014;**3**:e04501. DOI: 10.7554/eLife.04501

[54] Uhrig JF, Mutondo M, Zimmermann I, Deeks MJ, Machesky LM, Thomas P. The role of Arabidopsis scar genes in arp2-arp3-dependent cell morphogenesis. Development. 2007;**134**:967-977

[55] Makoto Y, Chunhua Z, Szymanski DB. Arp2/3-dependent growth in the plant kingdom: Scars for life. Frontiers in Plant Science. 2013;**4**:166. DOI: 10.3389/fpls.2013.00166

[56] Basu D, Le J, El Essal ED, Huang S, Zhang C, Mallery EL. Distorted3/scar2 is a putative Arabidopsis wave complex subunit that activates the arp2/3 complex and is required for epidermal morphogenesis. The Plant Cell. 2005;**17**:502-524

[57] Frank M, Egile C, Dyachok J, Frank M, et al. Activation of Arp2/3 complex-dependent actin polymerization by plant proteins distantly related to Scar/WAVE. Proceedings of the National Academy of Sciences. 2004;**101**:16379-16384

[58] Ali MF, Fatema U, Peng X, Hacker SW, Maruyama D, Sun MX, et al. ARP2/3-independent WAVE/SCAR pathway and class XI myosin control sperm nuclear migration in flowering plants. Proceedings of the National

Academy of Sciences of the United States of America. 2020;**51**:117

[59] Mogensen HL. Pollen tube-synergid interactions in Proboscidea louisianica (Martineaceae). American Journal of Botany. 1978;**65**:953-964. DOI: 10.2307/2442682

[60] Palanivelu R, Preuss D. Distinct short-range ovule signals attract or repel *Arabidopsis thaliana* pollen tubes in vitro. BMC Plant Biology. 2006;**6**:7. DOI: 10.1186/1471-2229-6-7

[61] Shimizu KK, Okada K. Attractive and repulsive interactions between female and male gametophytes in Arabidopsis pollen tube guidance. Development. 2000;**127**:4511-4518. DOI: 10.5167/uzh-71801

[62] Huck N, Moore JM, Federer M, Grossniklaus U. The Arabidopsis mutant feronia disrupts the female gametophytic control of pollen tube reception. Development. 2003;**130**:2149-2159. DOI: 10.1242/dev.00458

[63] Rotman N, Rozier F, Boavida L, Dumas C, Berger F, Faure JE. Female control of male gamete delivery during fertilization in *Arabidopsis thaliana*. Current Biology. 2003;**13**:432-436. DOI: 10.1016/S0960-9822(03)00093-9

[64] Escobar Restrepo J-M, Huck N, Kessler S, Gagliardini V, Ghey Selinck J, Yang W-C, et al. The FERONIA receptor-like kinase mediates male–female interactions during pollen tube reception. Science. 2007;**317**:656-660. DOI: 10.1126/science.1143562

[65] Capron A, Gourgues M, Neiva LS, Faure JE, Berger F, Pagnussat G, et al. Maternal control of male-gamete delivery in Arabidopsis involves a putative GPI-anchored protein encoded by the LORELEI gene. The Plant Cell. 2008;**20**:3038-3049. DOI: 10.1105/tpc.108.061713

[66] Kasahara RD, Maruyama D, Hamamura Y, Sakakibara T, Twell D, Higashiyama T. Fertilization recovery after defective sperm cell release in Arabidopsis. Current Biology. 2012;**22**:1084-1089. DOI: 10.1016/j.cub.2012.03.069

[67] Kasahara RD, Maruyama D, Higashiyama T. Fertilization recovery system is dependent on the number of pollen grains for efficient reproduction in plants. Plant Signaling & Behavior. 2013;**8**:e23690. DOI: 10.4161/psb.23690

[68] Zhong S, Li L, Wang Z, Ge Z, Li Q, Bleckmann A, et al. RALF peptide signaling controls the polytubey block in Arabidopsis. Science. 2022;**375**:290-296

[69] Duan Q, MCJ L, Kita D, et al. FERONIA controls pectin- and nitric oxide-mediated male–female interaction. Nature. 2021;**579**:561-566. DOI: 10.1038/s41586-020-2106-2

[70] Yu X, Zhang X, Zhao P, et al. Fertilized egg cells secrete endopeptidases to avoid polytubey. Nature. 2021;**592**:433-437. DOI: 10.1038/s41586-021-03387-5

[71] Kasahara RD, Notaguchi M, Nagahara S, Suzuki T, Susaki D, Honma Y, et al. Pollen tube contents initiate ovule enlargement and enhance seed coat development without fertilization. Science Advances. 2016;**2**:e1600554. DOI: 10.1126/sciadv.1600554

[72] Beeckman T, De Rycke R, Viane R, Inzé D. Histological study of seed coat development in *Arabidopsis thaliana*. Journal of Plant Research.

2000;**113**:139-148. DOI: 10.1007/pl00013924

[73] Figueiredo DD, Batista RA, Roszak PJ, Hennig L, Köhler C. Auxin production in the endosperm drives seed coat development in Arabidopsis. eLife. 2016;**5**:e20542. DOI: 10.7554/ eLife.20542

[74] Liu X, Adhikari PB, Kasahara RD. Pollen tube contents from failed fertilization contribute to seed coat initiation in Arabidopsis. F1000Research. 2019;**8**:348. DOI: 10.12688/f1000research.18644.2

[75] Liu X, Adhikari PB, Kasahara RD. Pollen tube content facilitates and increases the potential of endosperm proliferation irrespective of fertilization in *Arabidopsis thaliana*. F1000Research. 2019;**8**:348

[76] Honma Y, Adhikari PB, Kuwata K, Kagenishi T, Yokawa K, Notaguchi M, et al. High-quality sugar production by osgcs1 rice. Communications Biology. 2020;**3**:617. DOI: 10.1038/s42003-020-01329-x

Chapter 7

Taraxacum Kok-Saghys as a Strong Candidate Alternative Natural Rubber Crop in Temperate Regions in the Case of Emergency

Maryam Salehi, Moslem Bahmankar and Mohammad Reza Naghavi

Abstract

Natural rubber (NR, *cis*-1,4-polyisoprene) used in over 50,000 products, has unique properties, which cannot be matched by synthetic rubber. *Hevea brasiliensis* Muell. Arg. is currently the only NR commercial source that is not secure because of *Hevea* tree diseases, increasing demand, high labor costs, price instability, trade politics, competition for land with other crops, and a deforestation ban preventing new *H. brasiliensis* acreage. Hence, alternative rubber-producing crops are required for increasing the geographic and biological diversity of NR production. The mechanical properties and molecular composition of *Taraxacum kok-saghyz* NR are nearly identical to those of *H. brasiliensis* NR. However, developing *T. kok-saghyz* as an industrial crop is faced with some problems. This plant can become a commercially viable rubber-producing crop by improving agronomic fitness, rubber yield, and extraction process efficiency. An efficient process should extract NR at a high yield without damaging its physical and mechanical properties. This chapter focuses on the potential ways to improve rubber production and extraction processes from *T. kok-saghyz*.

Keywords: *cis*-polyisoprene, *Hevea brasiliensis*, molecular weight, rubber extraction, rubber yield, rubber quality

1. Introduction

Natural rubber (NR, *Cis*-1,4-polyisoprene) applied in over 50,000 products [1, 2] has unique properties which cannot be matched by synthetic rubber including abrasion and impact resistance, efficient heat dispersion, elasticity, malleability at cold temperatures, and resilience [1, 2]. *Trans*-polyisoprene has poorer low-temperature thermoplasticity (Tm = 54.5°C) and less flexibility compared with NR [1]. Synthetic rubbers are polymers of alkenes or dienes derived from petroleum which is a non-renewable resource [3]. *Trans*-polyisoprene possesses different characteristics which include high rigidity, very low coefficient of thermal contraction/expansion, outstanding insulation, and resistance to acid and alkali situations [1]. Due to its unique properties, this polymer is suitable for use in insulated cables, sporting goods, molds, dental products, and medical and scientific instruments.

H. brasiliensis Muell. Arg. is the only commercial NR source [1]. The amazon basin as a native region of *H. brasiliensis* only produces 2% of NR world production [4] because of South American leaf blight [a fatal endemic disease] on this continent. Southeast Asia countries, especially Thailand (4.85 million metric tons in 2019) and Indonesia (3.30 million metric tons in 2019) are the main NR producers [1, 2]. Attempts to develop *H. brasiliensis* clones having stable resistance and accepted yields have not been successful [1]. *H. brasiliensis* tree may enable to escape the disease by planting in escape or sub-optimal areas because long dry seasons interrupt the development cycle of fungus but yield decreases. The new clones are needed in such marginal and an agroforestry system can be planted with other perennial crops such as coffee or cocoa to achieve profitability [5, 6]. Currently, Global NR sources are not secure because of *Hevea* tree diseases, increasing demand, high labor costs, price instability, business politics, competition for land with other crops, and deforestation ban [1, 2]. Hence, alternative NR-producing plants are needed.

Many plants were studied in terms of rubber production potential when supply problems arose due to NR price or accessibility, particularly during World War I (the 1910s), World War II (the 1940s), and oil embargo (the 1970s). One of the prominent efforts is the establishment of the Edison Botanical Research Company in 1927. This company evaluated more than 17,000 rubber-producing plants in terms of NR content and quality [1].

2. Alternative rubber-producing plants to *H. brasiliensis*

About 2500 plant species produce *cis*-polyisoprene NR which is not always in a form of tappable latex [7]. It is noteworthy that molecular weight strongly correlates with quality and the most significant indication of usable rubber is a molecular weight of 1000 kg/mol or more [1, 2]. NR with an average molecular weight of over 1000 kg/mol is produced by only a few plants including para rubber tree (*H. brasiliensis*), *Parthenium argentatum* (guayule), *T. kok-saghyz* (rubber dandelion), *Taraxacum brevicorniculatum*, *Scorzonera tau-saghyz*, *Scorzonera uzbe-kistanica*, prickly lettuce (*Lactuca serriola*), lettuce (*Lactuca sativa*), *Ficus bengalensis*, and Madagascar rubber vine (*Cryptostegia grandiflora*) [1, 2].

The potential of Prickly lettuce as an alternative rubber source is not certain because of its low rubber content. *Scorzonera uzbekistanica*, *S. tau-saghyz*, and *T. kok-saghyz* can accumulate significant rubber content [1], but this takes years to accumulate.

Most *T. kok-saghyz* plants accumulate below 10% NR in the first cultivation year, but rare plants may accumulate above 20% NR [8]. The mechanical properties and molecular composition of *T. kok-saghyz* NR are nearly identical to those of *H. brasiliensis* NR, simplifying development, while *P. argentatum* rubber compounds need alternative formulations to meet product characteristics [9]. Many rubber companies are currently involved in *P. argentatum* and *T. kok-saghyz* rubber. However, it is not possible to go straight to a commodity tire market, so the success key is solving scalability issues.

T. kok-saghyz can be planted as an annual crop in temperate regions. Development of *T. kok-saghyz* as an industrial crop is faced with some problems including self-incompatibility, high heterozygosity, very variable rubber concentration, considerable rubber yield only quantifiable at maturity, its need for continual moisture during germination, low rate of growth, and poor competition with weeds [10, 11]. Hence, conventional and molecular breeding is required for improving agronomic fitness and NR yield to convert this species into a commercially reasonable rubber crop. Also, converting *T. kok-saghyz* into an industrial rubber-producing crop requires a cost-effective and environmentally friendly rubber extraction process proven at a commercial scale. This chapter focuses on the potential ways to improve rubber production and different latex and solid rubber extraction processes from *T. kok-saghyz*.

3. NR biosynthesis pathway

NR is composed of isopentenyl monomers derived from isopentenyl pyrophosphate, synthesized primarily from the cytosolic mevalonate pathway and likely also from the plastidic 2-C-methyl-D-erythritol-4-phosphate pathway (**Figure 1**) [1]. Geranyl pyro-phosphate, farnesyl pyrophosphate (FPP), and geranylgeranyl pyrophosphate can serve as rubber molecule initiators [12, 13]. FPP is most likely the leading *in vivo* initiator.

Rubber biosynthesis (**Figure 1**) is catalyzed by RT-ase (EC 2.5.1.20) at the rubber particle surface [14]. RT-ase is the only *cis*-prenyltransferase (CPT) that can biosynthesize high molecular weight *cis*-polyisoprene (>1000 kg/mol). Even *CPT*s that are not part of RT-ase, can affect rubber biosynthesis because CPTs are associated with sterol biosynthesis, and they are possibly required for forming rubber particle membranes. Also, most CPTs produce short-chain *cis*-allylic pyrophosphates which can function as rubber polymer initiators [15]. It seems that small rubber particle protein and rubber elongation factor are involved in NR biosynthesis [16–18] and they are effective targets for knockout or overexpression. A substantial mechanism regulating the secondary metabolite pathway is the transcriptional co-regulation of its pathway genes [19, 20]. Hence, this may demonstrate a main research scope in the future. Also, CPT-Like/CPT-binding proteins seem to be involved in NR biosynthesis [21–23] and may supply structural scaffolds [22].

4. Potential ways to improve NR yield

Theoretically, the reconstitution of rubber particles, a rubber synthetic machinery has not yet been achieved [1]. A eukaryotic organism with an endomembrane is required for ectopic rubber biosynthesis because eukaryotic post-translation modification may modify the RT-ase and also, the biogenesis of rubber particles likely occurs in Golgi or endoplasmic reticulum [24]. The overexpression of *hydroxymethylglutaryl coenzyme A reductase* [25] and allylic pyrophosphates [26] may increase rubber production, but it may harm rubber quality especially molecular weight [26]. It was noteworthy that the result of isoprenoid overexpression studies depends on the genetic background [1]. There is a negative correlation between rubber and inulin contents in *T. kok-saghyz* [14, 27, 28]. The target secondary metabolite content may be increased by blocking the branch pathways [29]. The content of short-chain polyisoprene was increased (two folds) in *T. kok-saghyz* and *T. brevicorniculatum* roots by over-expressing *fructan 1-exohydrolase*, which catalyzes the degradation of inulin to fructose and sucrose [30]. In brief, high molecular weight rubber biosynthesis in plants, yeasts,

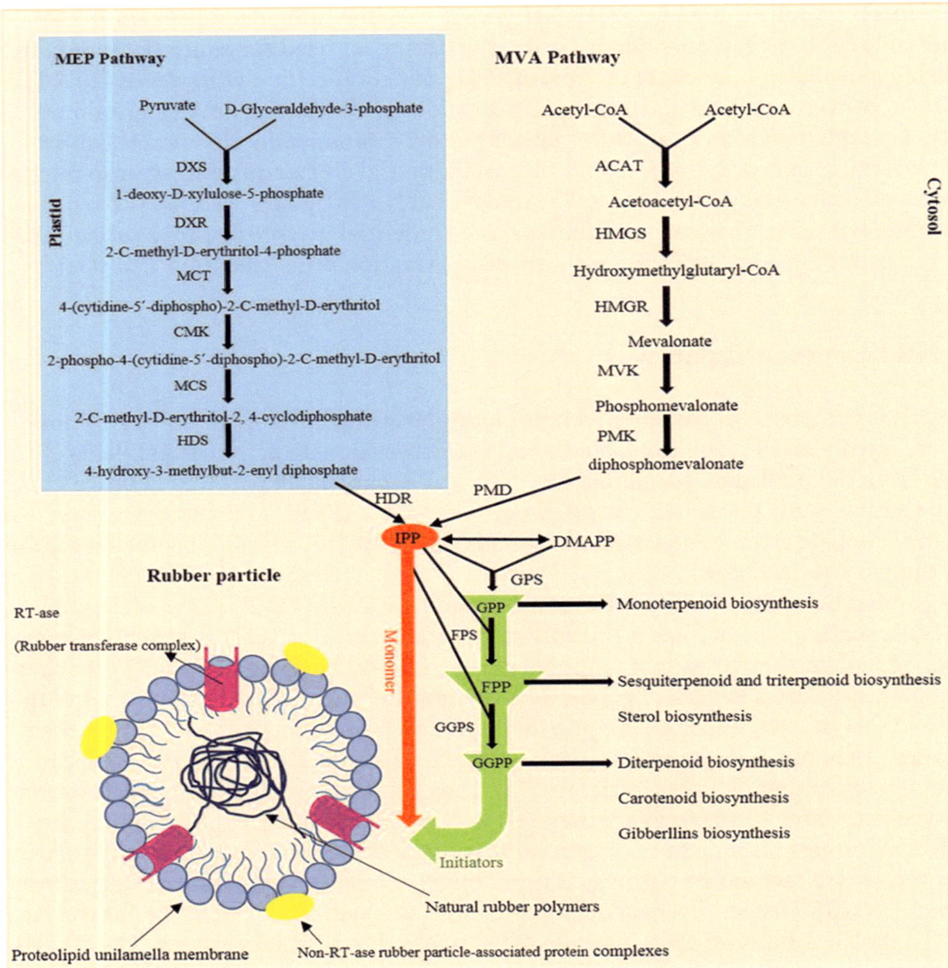

Figure 1.
The pathway of natural rubber (cis-1,4-poly isoprene), 2-C-methyl-D-erythritol-4-phosphate (MEP), mevalonate (MVA), and the oligomeric allylic pyrophosphate biosynthesis [1]. DXS: 1-deoxy-D-xylulose-5-phosphate synthase; DXR: 1-deoxy-D-xylulose-5-phosphate reductoisomerase; MCT: 2-C-methyl-D-erythritol-4-phosphate cytidyl transferase; CMK: 4-(cytidine-5′-diphospho)-2-C-methyl-D-erythritol kinase; MCS: 2-C-methyl-D-erythritol-2,4-cyclodiphosphate synthase; HDS: 4-hydroxy-3methylbut-2-enyl diphosphate synthase; HDR: 4-hydroxy-3-methylbut-2-enyl diphosphate reductase; ACAT: acetyl coenzyme A acetyltransferase; HMGS: hydroxymethylglutaryl coenzyme A synthase; HMGR: hydroxymethylglutaryl coenzyme A reductase; MVK: mevalonate kinase; PMK: phosphomevalonate kinase; PMD: diphosphomevalonate decarboxylase.

and bacteria has not been achieved by molecular genetic studies, and only isoprenoid production is slightly increased [14, 22, 30–34]. This indicates that our knowledge about feedback mechanisms and other rate-limiting enzymes is incomplete.

Breeding of *T. kok-saghyz* is essential in terms of agronomic properties, vigor traits, and rubber yield to convert it to a commercial crop. Hybridization [*Taraxacum officinale* (common dandelion) × *T. kok-saghyz*], selection, hybrid breeding, and polyploidy breeding have been assessed to achieve this aim. Quantitative traits can be selected by the family selection in the best way. For example, population selection and half-sib family selection increased the germination rate from 5.8% in cycle 0 to 40.8%

and 47.8% in cycle 3, respectively, under *in vitro* water stress [35]. Also, half-sib family, recurrent selection increased rubber yield from 0.15 to 0.22 g/plant after four selection cycles [36]. Interbreeding of diploid dandelion species failed to improve vigor and rubber concentration in single *Taraxacum* genotypes in early attempts [37]. However, recent European attempts have been successful to introgress vigor-related genetic elements of *T. officinale* into a rubber-producing genetic background of *T. kok-saghyz* [38]. The polyploidy breeding approach can enhance rubber concentration and/or root size and potentially end in high rubber yield. The self-incompatibility of *T. kok-saghyz* results in heterozygous and complex genetic backgrounds of its seeds, which may confound the chromosome doubling effects. Rare vigorous *T. kok-saghyz* tetraploids obtained in polyploid research [39, 40] propose that tetraploid breeding stands as a hopeful method for increasing *T. kok-saghyz* rubber yield. Developing the homozygous or inbred *T. kok-saghyz* lines are challenging because the self-incompatibility of diploid *T. kok-saghyz* prevents its self-fertilization. Development of *T. kok-saghyz* hybrids is possible using self-compatibility (SC) and cytoplasmic male sterility (CMS) traits [1]. CS and CMS have been detected [11] and identified [41] in *T. kok-saghyz*. Unfortunately, SC and CMS phenotypes are temperature-dependent [41].

Either hydroponic production or genetically modified *in vitro* cultures can be more rapidly scaled up than field crops and could provide essential, more expensive, NR in the case of emergency. Several strategies can enhance secondary metabolite production in *in vitro* cultures, for example, cell metabolism modification applying elicitors can result in increased secondary metabolite production in plant cells [42–45].

5. Rubber extraction processes

Rubber exists as latex (a rubber particle aqueous emulsion) and solid rubber threads in *T. kok-saghyz* living roots [2]. For extracting the rubber particles in latex form, *T. kok-saghyz* roots must be mechanically homogenized [46, 47]. The latex and solid rubber extraction processes date back to the 1930s. Soviet research focused on latex extraction [48–50] and US Researchers simulated mastication using wet-milling to extract solid rubber in water [51–54]. If the latex is not needed, it is better to recover all the NR as solid rubber by drying the roots [2].

For economic viability, all *T. kok-saghyz* components must be used [2]. Rubber, sugar syrups, soluble fiber, food, food and beverage ingredients, and biofuels can be generated from *T. kok-saghyz*. The leaves also can be applied in salads, teas, and tisanes [47]. Polar organic fractions contain useful compounds such as lubricants, cosmetic ingredients, insect pheromones, sealants, adhesives, surfactants, and emulsifiers [55].

5.1 Latex rubber extraction

Soviet researchers extracted latex from *T. kok-saghyz* roots using the flow method, in which thin circular cut roots are put in the extraction medium to extract the latex into a stabilizing buffer, then it is centrifuged to recover the latex [48–50]. USDA scientists developed the blender method to extract latex from *P. argentatum* at lab and pilot scales [56, 57]. This process can be applied to extract latex from *T. kok-saghyz* roots [2].

The latex colloidal stability can be assessed by zeta-potential measure [58]. The latex is likely coagulated when the emulsions go acidic [58]. The latex colloidal stabilization is ordinarily kept by adding hydroxides (most commonly, ammonium hydroxide [59], or ethanolamine (ETA) [58]. Some bases (like KOH or ammonia) and

ETA possess enough bactericidal activity to maintain latex for several months [58]. ETA is a better stabilizer and more "green" compared with ammonia or KOH [2].

5.2 Solid rubber extraction

Drying the roots recover all the rubber as solid rubber. Several processes have been patented to extract rubber from dried *T. kok-saghyz* roots [47, 52–55, 60–62] and some processes have pending applications. They include wet milling, enzyme digestion, solvent extraction, and dry milling processes.

5.2.1 Wet-milling process

Wet milling simulates mastication using pebble milling to extract rubber in water [9, 51–54, 63]. In the Eskew process, at first, inulin and other water-soluble components are extracted from dried and chopped *T. kok-saghyz* roots by mixing the roots with hot water, then rubber is separated from root biomass using pebble milling, and in the end, the rubber is retrieved by flotation [51–54]. No enzymes or chemicals are needed in the Eskew process (wet milling) [52], so it is a low-cost process [2]. The rubber purity reduces in this method because root biomass is trapped in the rubber phase [2]. Eskew process can retrieve nearly 75% of the rubber in *T. kok-saghyz* roots with a dirt content of 12–15% w/w dry NR [53], which hurts rubber mechanical characteristics and restricts its commercial application. Two processes were patented for improving the yield and purity of rubber extracted by the Eskew process. One of them reported that the rubber yield was increased to 90% by milling whole roots instead of chopped roots [52], but the rubber impurity was still 10–15% [53]. In another patent, the crude rubber extracted by Eskew process was scrubbed in NaOH solution, and then it was neutralized with stearic acid to improve rubber purity (1.47% impurities) [53].

5.2.2 Enzyme digestion process

About 77% (w/w) of rubber impurities extracted using the Eskew process include cellulose, hemicellulose, lignin, and pectin [64]. So, in the enzymatic digestion process, after carbohydrate extraction from dried and crushed roots (less than 1 mm), rubber is further purified by a mixture of industrial cellulose, hemicellulose, and xylanase enzymes [47, 61, 64].

The PENRA III, an enzyme-based aqueous process was developed by the team of PENRA (Program for Excellence in Natural Rubber Alternatives). This process consists of extracting with hot water, alkaline pre-treatment, treating with enzyme, centrifuging, pebble milling, floating, and filtrating [63]. The rubber yield and purity were 80% (of the theoretical *T. kok-saghyz* NR yield) [64] and 99.5% (at lab-scale) [65], respectively in this process. The enzymes secreted from *Thermomyces lanuginosus* fungus and enzymes contained in transgenic maize flour may be used as low-cost enzymes in the commercial extraction process [2].

5.2.3 Solvent extraction

In solvent extraction, inulin is preliminary extracted from dried and ground *T. kok-saghyz* roots using hot water, then polar and non-polar solvents are sequentially or simultaneously used [66] for extracting and purifying rubber. The purity requirements of 99.8% are met in solvent extraction [67] but may lead to safety and environmental concerns. Also, it is faced with scale-up challenges and is expensive. Also, this process may result in lower rubber yield because *T. kok-saghyz* rubber has naturally crosslinked rubber gel which difficulty dissolves in solvents [68].

5.2.4 Dry milling process

T. kok-saghyz roots dried to 7.5% moisture were ground by a gristmill, then rubber threads were separated from root tissue applying a mesh screen (mesh size: 2 mm). The rubber threads were shaken in warm water, settled, floating rubber threads, skimmed off, and re-stirred in warm water [60]. This process can recover 97.5% of the extractable rubber with a purity of 99.8% after 5 repetitions.

5.3 Natural rubber quality

On the molecular level, the determinants of rubber quality consist of polymer molecular weight, macromolecular structure (branching), gel content, and content and composition of non-rubber components like lipids and proteins [1, 2]. Hence, the extracted rubber should be assessed in terms of rubber purity, gel content of the solid rubber, resin content, molecular characterization, and NR composition. An impurity content of less than 0.2% is required for preventing unallowable tear initiation and propagation according to ASTM D1278-91a [69].

6. Conclusions and perspectives

Trying to establish *H. brasiliensis* clones having acceptable yield and durable resistance has not been successful. *T. kok-saghyz* as a strong candidate alternative rubber crop needs optimized agronomic practices and extraction processes. New *T. kok-saghyz* plants can be developed using gene editing and breeding research. None of the studied processes for *T. kok-saghyz* rubber extraction are fully satisfactory and each of them has intrinsic advantages and disadvantages. It is necessary to optimize the extraction process that can produce rubber with high purity and high quality, is cost-effective, and does not cause environmental concerns.

Conflict of interest

The authors declare no conflict of interest.

Abbreviations

NR	natural rubber
FPP	Farnesyl pyrophosphate
CPT	*cis*-prenyltransferase
SC	self-compatibility
CMS	cytoplasmic male sterility
ETA	Ethanolamine
PENRA	program for excellence in natural rubber alternatives

Author details

Maryam Salehi[1*], Moslem Bahmankar[2*] and Mohammad Reza Naghavi[1]

1 Division of Biotechnology, Department of Agronomy and Plant Breeding, College of Agricultural and Natural Resources, University of Tehran, Karaj, Iran

2 Department of Seed and Plant Improvement Research, Safiabad Agricultural Research, Education and Natural Resources Center, AREEO, Dezful, Iran

*Address all correspondence to: salehi_maryam_83@yahoo.com and m.bahmankar@areeo.ac.ir

IntechOpen

© 2023 The Author(s). Licensee IntechOpen. This chapter is distributed under the terms of the Creative Commons Attribution License (http://creativecommons.org/licenses/by/3.0), which permits unrestricted use, distribution, and reproduction in any medium, provided the original work is properly cited.

References

[1] Salehi M, Cornish K, Bahmankar M, Naghavi MR. Natural rubber-producing sources, systems, and perspectives for breeding and biotechnology studies of *Taraxacum kok-saghyz*. Industrial Crops and Products. 2021;**170**:113667. DOI: 10.1016/j.indcrop.2021.113667

[2] Salehi M, Bahmankar M, Naghavi MR, Cornish K. Rubber and latex extraction processes for *Taraxacum kok-saghyz*. Industrial Crops and Products. 2022;**178**:114562. DOI: 10.1016/j.indcrop.2022.114562

[3] Hirsch RL, Bezdek R, Wendling R. Peaking of world oil production and its mitigation. AIChE Journal. 2006;**52**:2-8. DOI: 10.1002/aic.10747

[4] Vaysse L, Bonfils F, Sainte-Beuve J, Cartault M. 10.1. 7-Natural rubber. In: McGrath JE, Hickner MA, Hofer R, editors. Polymer Science: A Comprehensive Reference. Polymers for a Sustainable Environment and Green Energy. Amsterdam: Elsevier; 2012. pp. 281-293

[5] Priyadarshan PM, Hoa TTT, Huasun H, de Goncalves PS. Yielding potential of rubber (*Hevea brasiliensis*) in sub-optimal environments. Journal of Crop Improvement. 2005;**14**:221-247. DOI: 10.1300/J411v14n01_10

[6] Snoeck D, Lacote R, Keli ZJ, Doumbia A, Chapuset T, Jagoret P, et al. Association of hevea with other tree crops can be more profitable than *Hevea monocrop* during first 12 years. Industrial Crops and Products. 2013;**43**:578-586. DOI: 10.1016/j.indcrop.2012.07.053

[7] Bonner J. The history of rubber. In: Whitworth JW, Whitehead EE, editors. Guayule Natural Rubber: A Technical Publication with Emphasis on Recent Findings. Guayule Administrative Management Committee and US Department of Agriculture Cooperative State Research Service. Tucson, AZ: Office of Arid Lands Studies, University of Arizona; 1991. pp. 1-16

[8] Cornish K, Kopicky SL, McNulty SK, Amstutz N, Chanon AM, Walker S, et al. Temporal diversity of *Taraxacum kok-saghyz* plants reveals high rubber yield phenotypes. Biodiversitas. 2016;**17**:847-856. DOI: 10.13057/biodiv/d170262

[9] Ramirez-Cadavid DA, Cornish K, Michel FC Jr. Alternative NR production: An investigation has been carried out into the processing of *Taraxacum kok-saghyz* roots to produce natural rubber (NR), inulin and co-products. Tire Technology International Annual Review. 2020;**2020**:60-63

[10] Warmke HE. Macrosporogenesis, fertilization, and early embryology of *Taraxacum kok-saghyz*. Bulletin of the Torrey Botanical Club. 1943;**70**:164-173. DOI: 10.2307/2481367

[11] Hodgson-Kratky KJM, Wolyn DJ. Cytoplasmic male sterility in Russian dandelion. Journal of the American Society for Horticultural Science. 2015;**140**:580-586. DOI: 10.21273/JASHS.140.6.580

[12] Gutensohn M, Orlova I, Nguyen TT, Davidovich-Rikanati R, Ferruzzi MG, Sitrit Y, et al. Cytosolic monoterpene biosynthesis is supported by plastid-generated geranyl diphosphate substrate in transgenic tomato fruits. The Plant Journal. 2013;**75**:351-363. DOI: 10.1111/tpj.12212

[13] Mendoza-Poudereux I, Kutzner E, Huber C, Segura J, Eisenreich W,

Arrillaga I. Metabolic cross-talk between pathways of terpenoid backbone biosynthesis in spike lavender. Plant Physiology and Biochemistry. 2015;**95**:113-120. DOI: 10.1016/j.plaphy.2015.07.029

[14] Cornish K, Xie W. Natural rubber biosynthesis in plants: Rubber transferase. Methods in Enzymology. 2012;**515**:63-82. DOI: 10.1016/B978-0-12-394290-6.00004-5

[15] Cornish K, Siler DJ, Grosjean OK, Goodman N. Fundamental similarities in rubber particle architecture and function in three evolutionarily divergent plant species. Journal of Natural Rubber Research. 1993;**8**:275-285

[16] Dennis MS, Light DR. Rubber elongation factor from *Hevea brasiliensis*. Identification, characterization, and role in rubber biosynthesis. Journal of Biological Chemistry. 1989;**264**:18608-18617. DOI: 10.1016/S0021-9258(18)51510-6

[17] Oh SK, Kang H, Shin DH, Yang J, Chow KS, Yeang HY, et al. Isolation, characterization, and functional analysis of a novel cDNA clone encoding a small rubber particle protein from *Hevea brasiliensis*. Journal of Biological Chemistry. 1999;**274**:17132-17138. DOI: 10.1074/jbc.274.24.17132

[18] Kim IJ, Ryu SB, Kwak YS, Kang H. A novel cDNA from *Parthenium argentatum* Gray enhances the rubber biosynthetic activity *in vitro*. Journal of experimental botany. 2004;**55**:377-385. DOI: 10.1093/jxb/erh039

[19] Salehi M, Karimzadeh G, Naghavi MR, Naghdi Badi H, Rashidi MS. Expression of artemisinin biosynthesis and trichome formation genes in five *Artemisia* species. Industrial Crops and Products. 2018;**112**:130-140. DOI: 10.1016/j. indcrop.2017.11.002

[20] Bahmankar M, Mortazavian SMM, Tohidfar M, Noori SAS, Izadi Darbandi A, Salehi M, et al. Physio-biochemical characters, embryo regeneration and limonene synthase gene expression in cumin. Industrial Crops and Products. 2018;**121**:195-205. DOI: 10.1016/j.indcrop.2018.05.006

[21] Brasher MI, Surmacz L, Leong B, Pitcher J, Swiezewska E, Pichersky E, et al. A two-component enzyme complex is required for dolichol biosynthesis in tomato. The Plant Journal. 2015;**82**:903-914. DOI: 10.1111/tpj.12859

[22] Qu Y, Chakrabarty R, Tran HT, Kwon EJG, Kwon M, Nguyen TD, et al. A lettuce (*Lactuca sativa*) homolog of human Nogo-B receptor interacts with *cis*-prenyltransferase and is necessary for natural rubber biosynthesis. Journal of Biological Chemistry. 2015;**290**:1898-1914. DOI: 10.1074/jbc.M114.616920

[23] Brown D, Feeney M, Ahmadi M, Lonoce C, Sajari R, Di Cola A, et al. Subcellular localization and interactions among rubber particle proteins from *Hevea brasiliensis*. Journal of Experimental Botany. 2017;**68**:5045-5055. DOI: 10.1093/jxb/erx331

[24] Men X, Wang F, Chen GQ, Zhang HB, Xian M. Biosynthesis of natural rubber: Current state and perspectives. International Journal of Molecular Sciences. 2019;**20**:50-71. DOI: 10.3390/ijms20010050

[25] Dong N, Ponciano G, McMahan CM, Coffelt TA, Johnson L, Creelman R, et al. Overexpression of 3-hydroxy-3-methylglutaryl coenzyme a reductase in *Parthenium argentatum* (guayule). Industrial Crops and Products. 2013;**46**:15-24. DOI: 10.1016/j.indcrop.2012.12.044

[26] Veatch ME, Ray DT, Mau CJD, Cornish K. Growth, rubber, and resin

evaluation of two-year-old transgenic guayule. Industrial Crops and Products. 2005;**22**:65-74. DOI: 10.1016/j.indcrop.2004.06.007

[27] Arias M, Herrero J, Ricobaraza M, Hernández M, Ritter E. Evaluation of root biomass, rubber and inulin contents in nine *Taraxacum koksaghyz* Rodin populations. Industrial Crops and Products. 2016;**83**:316-321. DOI: 10.1016/j.indcrop.2016.01.023

[28] Kreuzberger M, Hahn T, Zibek S, Schiemann J, Thiele K. Seasonal pattern of biomass and rubber and inulin of wild Russian dandelion (*Taraxacum koksaghyz* L. Rodin) under experimental field conditions. European Journal of Agronomy. 2016;**80**:66-77. DOI: 10.1016/j.eja.2016.06.011

[29] Salehi M, Karimzadeh G, Naghavi MR, Naghdi Badi H, Rashidi MS. Expression of key genes affecting artemisinin content in five *Artemisia* species. Scientific Reports. 2018b;**8**:12659. DOI: 10.1038/s41598-018-31079-0

[30] Stolze A, Wanke A, van Deenen N, Geyer R, Prüfer D, Schulze GC. Development of rubber-enriched dandelion varieties by metabolic engineering of the inulin pathway. Plant Biotechnology Journal. 2017;**15**:740-753. DOI: 10.1111/pbi.12672

[31] Schmidt T, Lenders M, Hillebrand A, van Deenen N, Munt O, Reichelt R, et al. Characterization of rubber particles and rubber chain elongation in *Taraxacum koksaghyz*. BMC Biochemistry. 2010;**11**:11-21. DOI: 10.1186/1471-2091-11-11

[32] Epping J, van Deenen N, Niephaus E, Stolze A, Fricke J, Huber C, et al. A rubber transferase activator is necessary for natural rubber biosynthesis in dandelion. Nature Plants. 2015;**1**:1-9. DOI: 10.1038/nplants.2015.48

[33] Kwon M, Kwon EJ, Ro DK. *Cis*-Prenyltransferase and polymer analysis from a natural rubber perspective. Methods in Enzymology. 2016;**576**:121-145. DOI: 10.1016/bs.mie.2016.02.026

[34] Pütter KM, van Deenen N, Unland K, Prüfer D, Gronover CS. Isoprenoid biosynthesis in dandelion latex is enhanced by the overexpression of three key enzymes involved in the mevalonate pathway. BMC Plant Biology. 2017;**17**:88-100. DOI: 10.1186/s12870-017-1036-0

[35] Hodgson-Kratky KJ, Stoffyn OM, Wolyn DJ. Recurrent selection for improved germination under water stress in Russian Dandelion. Journal of the American Society for Horticultural Science. 2017;**142**:85-91. DOI: 10.21273/JASHS03941-16

[36] Hodgson-Kratky KJ, Stoffyn OM, Wolyn DJ. Recurrent selection for rubber yield in Russian Dandelion. Journal of the American Society for Horticultural Science. 2017b;**142**:470-475. DOI: 10.21273/JASHS04252-17

[37] Krotkov G. A review of literature on *Taraxacum koksaghyz* rod. The Botanical Review. 1945;**11**:417-461. DOI: 10.1007/BF02861139

[38] van Dijk PJ, Sørensen AP. Rubber producing *Taraxacum* plant. 2018; US20180271043

[39] Warmke HE. Experimental polyploidy and rubber content in *Taraxacum kok-saghyz*. Botanical Gazette. 1945;**106**:316-324. DOI: 10.1086/335301

[40] Luo Z, Iaffaldano BJ, Cornish K. Colchicine-induced polyploidy has the potential to improve rubber yield in *Taraxacum kok-saghyz*. Industrial Crops and Products. 2018;**112**:75-81. DOI: 10.1016/j.indcrop.2017.11.010

[41] Di T. Genetic analysis of male sterility and self-compatibility in Russian Dandelion (*Taraxacum kok-saghyz*). Doctoral Dissertation. 2017

[42] Salehi M, Karimzadeh G, Naghavi MR. Synergistic effect of coronatine and sorbitol on artemisinin production in cell suspension culture of *Artemisia annua* L. cv Anamed. Plant Cell, Tissue and Organ Culture. 2019;**137**:587-597. DOI: 10.1007/s11240-019-01593-01598

[43] Salehi M, Naghavi MR, Bahmankar M. A review of *Ferula* species: Biochemical characteristics, pharmaceutical and industrial applications, and suggestions for biotechnologists. Industrial Crops and Products. 2019;**139**:111511-111519. DOI: 10.1016/j.indcrop.2019.111511

[44] Salehi M, Farhadi S, Moieni A, Safaie N, Ahmadi H. Mathematical modeling of growth and paclitaxel biosynthesis in *Corylus avellana* cell culture responding to fungal elicitors using multilayer perceptron-genetic algorithm. Frontiers in Plant Science. 2020;**11**:1148-1159. DOI: 10.3389/fpls.2020.01148

[45] Salehi M, Moieni A, Safaie N, Farhadi S. Whole fungal elicitors boost paclitaxel biosynthesis induction in *Corylus avellana* cell culture. PLoS One. 2020b;**15**:e0236191. DOI: 10.1371/journal.pone.0236191

[46] Cornish K, Xie W, Kostyal D, Shintani D, Hamilton RG. Immunological analysis of the alternate rubber crop *Taraxacum kok-saghyz* indicates multiple proteins cross-reactive with *Hevea brasiliensis* latex allergens. Journal of Biotechnology and Biomaterials. 2015;**5**:201-207. DOI: 10.4172/2155-952X.1000207

[47] Swiger D, Wade J. Dandelion processes, compositions and products. 2019. U.S. Patent No. 10,184,029

[48] Ignat'ev AM. Vozmozhnosti polucheniia natural'nogo lateksa iz kauchukonosov [On the feasibility of obtaining natural latex from the rubber-bearing plants]. Soviet Kauchuk. 1935;**3**:22-26

[49] Ignat'ev AM. Poluchenie lateksa iz sagyzov [extraction of latex from various saghyzes]. *Zhurnal Rezinovoi Promyshlennosti*. 1936;**3**:1070-1078

[50] Ignat'ev AM, Uzina K, Efrofeev T. Khranenie kornei koksagyza I poluchenie lateksa [Storage of kok-saghyz and extraction of latex]. *Kauchuk i Rezina*. 1940;**1**:30-33

[51] Eskew RK. Natural rubber from Russian Dandelion. Rubber Chemistry and Technology. 1946;**19**:856-864. DOI: 10.5254/1.3543236

[52] Eskew RK, Edwards PW. Process for recovering rubber from fleshy plants. 1946. US Patent US2393035 A

[53] Stamberger P, Eskew RK, Hanslick RS. Treatment of rubber. 1946. US Patent US2399156 A

[54] Whaley GW, Bowen JS. Russian Dandelion (Kok-Saghyz). An Emergency Source of Natural Rubber; USDA Miscellaneous Publication 618. Washington, DC: U.S. Government Printing Office; 1947. p. 210

[55] Nocera A Jr, Swiger DR. Rubber and by-product extraction systems and methods. 2020. U.S. Patent 10,584,185

[56] Cornish K. Hypoallergenic natural rubber products from *Parthenum argentatum* (gray) and other non-*Hevea brasiliensis* species. 1996. U.S. Patent No. 5,580,942

[57] Cornish K. Hypoallergenic natural rubber products from *Parthenium argentatum* (Gray) and other non-*Hevea brasiliensis* species. 1998. U.S. Patent No. 5,717,050

[58] Krickl S, Touraud D, Kunz W. Investigation of ethanolamine stabilized natural rubber latex from *Taraxacum kok-saghyz* and from *Hevea brasiliensis* using zeta-potential and dynamic light scattering measurements. Industrial Crops and Products. 2017;**103**:169-174. DOI: 10.1016/j.indcrop.2017.03.046

[59] Blackley DC. Polymer Latices. 2nd ed. Netherlands, Dordrecht: Springer; 1997. DOI: 10.1007/978-94-011-5866-4

[60] Buranov AU. Process for recovering rubber from rubber-bearing plants with a gristmill. 2009. U.S. Patent No. 7,540,438

[61] Wade J, Swiger D. Dandelion processes, compositions and products. 2016. U.S. Pat. 9 (346), 924

[62] Swiger D, Wade J. Dandelion processes, compositions and products. 2017. U.S. Patent No. 9,611,363

[63] Ramirez-Cadavid DA, Cornish K, Hathwaik U, Kozak R, McMahan C, Michel FC Jr. Development of novel processes for the aqueous extraction of natural rubber from *Taraxacum kok-saghyz* (TK). Journal of Chemical Technology and Biotechnology. 2019;**94**:2452-2464. DOI: 10.1002/jctb.6027

[64] Ramirez-Cadavid DA, Cornish K, Michel FC Jr. *Taraxacum kok-saghyz* (TK): Compositional analysis of a feedstock for natural rubber and other bioproducts. Industrial Crops and Products. 2017;**107**:624-640. DOI: 10.1016/j.indcrop.2017.05.043

[65] Liu P. Enhancing TK rubber extraction efficiency with fungus and enzyme treatments. [Doctoral dissertation], The Ohio State University; 2018

[66] Randall AM, Gozen AO. Compositions containing purified non-Hevea rubber and related purification methods. U.S. Patent No. 10,023,660. U.S. Patent and Trademark Office, Washington, DC. 2018

[67] Huang Y, Mouri H, Beaulieu M. Processes for the removal of rubber from TKS plant matter. 2015. US Patent App US 15/0073113 A1. USPTO

[68] Cornish K. Alternative natural rubber crops: Why should we care? Technology and Innovation. 2017;**18**: 244-255. DOI: 10.21300/18.4.2017.245

[69] Baranwal KC, Stephens HL. Basic Elastomer Technology Rubber Division. Akron, OH: American Chemical Society; 2001